# The Block Universe

## Exploring a Matrix Reality

Greg Rusk

Copyright © Greg Rusk, 2024

All rights reserved.

ISBN: 9798303484337

# Foreword

What is the universe? What lies beyond its boundaries? And what if spacetime itself is entirely different from how it is initially perceived by humans? For centuries, time has been central to how we make sense of reality, yet it remains one of the greatest enigmas. The deeper science and philosophy probe into its nature, the more complex and elusive it becomes.

This book grew out of a simple but persistent question: could our understanding of the universe be the key to unraveling some of modern physics' most profound mysteries? It explores the concept of the block universe, a framework that challenges the idea of time as a sequence of events flowing from past to future. Instead, it proposes that all moments, past, present, and future, exist simultaneously within an unchanging spacetime structure.

At first glance, this idea might feel counterintuitive, perhaps even unsettling. Yet it offers an opportunity to reexamine the nature of existence from a new perspective. The Block Universe presents a way to address one of physics' enduring challenges: reconciling Einstein's deterministic universe of relativity with the probabilistic world of quantum mechanics. Could this framework help bridge the gap between these two pillars of modern science?

This book delves into the scientific foundations of these ideas, their philosophical implications, and their potential to reshape how we see ourselves in the cosmos. While it is grounded in scientific inquiry, the ideas here extend beyond equations and theories. They touch on fundamental questions about reality, perception, and our search for meaning.

Not only is the Universe stranger than we think...
It's stranger than we *can* think.

-Werner Heisenberg

# Table of Contents

Introduction

Chapter 1: The Emergence of the Block Universe Hypothesis

Chapter 2: Relativity and the Foundations of the Block Universe

Chapter 3: Reconciling the Block Universe with Quantum Mechanics

Chapter 4: Philosophical and Metaphysical Underpinnings of Eternalism

Chapter 5: Observational Evidence and Theoretical Support

Chapter 6: Mechanisms for Navigating the Block (Theoretical)

Chapter 7: The Science of Closed Timelike Curves and Wormholes

Chapter 8: Quantum Interpretations and the Many-Worlds Conundrum

Chapter 9: Technological Challenges and the Quest for Exotic Matter

Chapter 10: Energy Requirements and Engineering the Impossible

Chapter 11: Ethical, Moral, and Existential Considerations

Chapter 12: Covert Applications: Government and Military Interests

Chapter 13: Unidentified Aerial Phenomena and Temporal Anomalies

Chapter 14: Integrating Consciousness Studies and Spacetime

Chapter 15: The Simulation Hypothesis and the Block Universe

Chapter 16: The Road Ahead: Future Research Directions

Chapter 17: Conclusion and the Grand Speculative Leap

References

## Introduction

Time is a concept that permeates every aspect of human life, yet its true nature remains one of the most profound mysteries in science and philosophy. For centuries, it has been approached as a flowing entity, carrying moments from the past into the present and onward to the future. But what if this familiar perception is misleading? What if time is not a flow but a structure, a timeless framework where every event, past or future, already exists?

This book explores the block universe hypothesis, a perspective that redefines time as part of a four-dimensional spacetime continuum. Unlike the linear progression of moments we experience, this view suggests that all events coexist within a unified and unchanging reality. For many, this notion feels counterintuitive, but it offers a way to reconcile some of the most puzzling questions in physics.

Modern theories of relativity and quantum mechanics are both remarkably successful at describing the universe but differ fundamentally in how they approach time. The block universe proposes a framework that challenges these contradictions, providing a deeper understanding of their implications. From Einstein's relativity to Minkowski's spacetime geometry, these ideas lay the groundwork for a view of the cosmos that defies conventional wisdom.

The implications of such a perspective stretch beyond equations and theories. How might the block universe shape our understanding of causality? What does it mean for concepts like free will, change, and the meaning of existence? These are questions that not only concern scientists but resonate with anyone interested in the nature of reality.

This book does not claim to provide all the answers. Instead, it presents a perspective grounded in science but open to the vast possibilities that lie beyond current understanding. Whether examining the deterministic underpinnings of relativity, the probabilistic nature of quantum mechanics, or even the tantalizing mysteries of phenomena yet unexplained, the block universe invites fresh inquiry.

By considering time as a dimension rather than a flow, this perspective offers a new way of seeing the cosmos, one that is as challenging as it is intriguing. For those curious about the nature of existence, it provides a lens to question the familiar and explore what may lie beyond.

# Chapter 1

## The Emergence of the Block Universe Hypothesis

The block universe hypothesis has emerged as a revolutionary way of thinking about time and reality. Unlike the traditional view of time as a flowing sequence of moments, this concept suggests that all moments coexist in a four-dimensional spacetime. This idea challenges how we perceive past, present, and future, introducing a paradigm where all points in time are equally real. In this chapter, we will trace the origins of the block universe hypothesis, explore its scientific underpinnings, and delve into its philosophical implications. By examining contributions from Einstein, Gödel, Wheeler, Feynman, and Hawking, we can see how this theory bridges relativity and quantum mechanics, reshaping our understanding of existence.

### Historical Steps Toward the Concept

For centuries, time was thought to be absolute. Isaac Newton's vision of time portrayed it as a constant, flowing backdrop against which the events of the universe played out. Time was unchanging, universal, and independent of matter or motion. This classical view dominated scientific thought until the early 20th century when Albert Einstein revolutionized our understanding of the cosmos.

Einstein's theory of relativity introduced the idea that time is not a separate, absolute entity but is intertwined with space, forming a four-dimensional spacetime continuum. In this framework, time is relative, influenced by factors such as gravity and velocity. For example, clocks run slower near massive objects or at high speeds, a phenomenon experimentally confirmed through satellite systems and atomic clocks. Einstein's insights laid the foundation for a new way of understanding time, one that would eventually lead to the block universe hypothesis.

This shift from Newton's absolute time to Einstein's relative time marked a profound turning point. No longer was time an independent constant; instead, it became a dynamic participant in the fabric of the universe. Events were no longer isolated occurrences but fixed points within a continuum, challenging humanity's perception of time as a flowing river.

**Gödel's Closed Timelike Curves**

Building on Einstein's equations, Kurt Gödel proposed solutions that introduced the concept of closed timelike curves. These theoretical constructs suggested that under specific conditions, it might be possible to traverse through time. Gödel's work was groundbreaking, as it implied that the universe could be structured in a way where the past, present, and future were interconnected in a single, cohesive framework.

Gödel's closed timelike curves not only reinforced the block universe hypothesis but also sparked debates about causality and free will. If time travel were possible, what would it mean for the concept of cause and effect? Could an individual influence events in the past, or would their actions be constrained by the predetermined structure of spacetime? These questions remain central to philosophical discussions about the nature of time and the implications of the block universe.

Gödel's work demonstrated that the mathematical framework of Einstein's relativity could accommodate scenarios far beyond our everyday experience. While time travel remains speculative, the possibility of closed timelike curves invites us to rethink the linear progression of time and consider a reality where all moments are accessible, though not necessarily alterable.

**Visualizing the Four-Dimensional Landscape**

John Wheeler and Richard Feynman's contributions to quantum mechanics introduced new dimensions to the discussion of time and causality. Their absorber theory proposed a symmetry between the past and future, where particles could influence each other across time. This idea challenged the traditional view of causality as a one-way street and suggested that events in the future could shape conditions in the past.

This quantum perspective aligns with the block universe, where all points in time coexist. Wheeler and Feynman's work highlighted the interconnectedness of events across spacetime, emphasizing that causality might not be as straightforward as it seems. Instead of thinking of time as a strict sequence of causes and effects, their theories suggest a web of relationships that transcend temporal boundaries.

Wheeler and Feynman's insights also raise intriguing questions about the role of observation in shaping reality. In quantum mechanics, the act of observation collapses a particle's wavefunction, determining its state. Within the block universe, observation could be seen as the mechanism by which consciousness navigates the spacetime continuum, selecting specific moments from a timeless landscape.

**Philosophical Roots and the Eternalism View**

Stephen Hawking's exploration of black holes and singularities provided additional support for the block universe hypothesis. Black holes, with their immense gravitational pull, warp spacetime to such an extent that traditional notions of time break down. Near a black hole, time slows dramatically, and at the event horizon, it effectively stops. This phenomenon demonstrates the malleability of time and reinforces the idea that it is not an absolute flow but a dimension subject to the fabric of spacetime.

Hawking's work also introduced the concept of Hawking radiation, where particles escape a black hole's grasp. This discovery bridged quantum mechanics and general relativity, two fields that often seem at odds. In the context of the block universe, Hawking's findings illustrate how extreme environments can reveal the intricate interplay between time, space, and matter, providing a glimpse into the fundamental nature of reality.

**Setting the Stage for Further Inquiry**

The block universe raises profound philosophical questions about free will and determinism. If all moments in time coexist, are our actions predetermined? Or do we possess the agency to shape our destinies within the fixed structure of spacetime? Philosophers have long debated whether free will is compatible with a deterministic universe, and the block universe adds new dimensions to this discussion.

One perspective is that free will is an emergent phenomenon, a subjective experience arising from our navigation of spacetime. While the timeline itself may be fixed, our perception of choice and decision-making provides a sense of agency. From this view, free will is not about altering the timeline but about how we experience and interpret our journey through it.

**Summary**

The emergence of the block universe hypothesis represents a paradigm shift in how we understand time and reality. From Einstein's relativity to Gödel's closed timelike curves, and from Wheeler and Feynman's quantum insights to Hawking's exploration of black holes, this concept integrates the most profound discoveries of modern science. By challenging the traditional view of time as a linear progression, the block universe invites us to reconsider our place in the cosmos and the nature of existence itself. This chapter sets the stage for deeper explorations into the scientific and philosophical implications of a timeless reality.

## Chapter 2

## **Relativity and the Foundations of the Block Universe**

In the early decades of the 20th century, as physicists grappled with increasingly precise experiments and unusual findings, an entirely new understanding of time and space began to emerge. This understanding challenged centuries of accepted thought and philosophical comfort. Central to this shift was the theory of relativity, formulated by Albert Einstein in a series of papers beginning in 1905. Prior to Einstein's revolutionary work, most scientists and scholars still operated under the assumption that time was absolute and universal, and that space was a fixed stage upon which events took place. Einstein's insights not only dismantled these long held beliefs, they prepared the ground for what we now call the block universe hypothesis.

The block universe idea proposes that all moments in time coexist in a four dimensional structure called spacetime. Rather than picturing reality as evolving through a single universal now that changes as time flows forward, the block universe presents a scenario in which past, present, and future all have equal footing as points within a timeless continuum. In essence, this viewpoint treats time as a dimension similar to the three familiar dimensions of space. Although Einstein himself did not explicitly use the term block universe, his relativistic framework led directly to interpretations and philosophies that strongly support this concept.

As we move through this chapter, we will see how the shift from Newton's absolute time to Einstein's relative time paved the way for a more geometric and less time bound view of reality. While we will keep the tone human and occasionally add a small humorous remark to maintain engagement, the focus here will be on the historical developments, scientific details, and the conceptual arguments that transformed our understanding of time.

## From Newton to Einstein – The Collapse of Absolute Time

For a large part of modern history, Isaac Newton's views on time and space held tremendous sway. In his 1687 work, the Principia Mathematica, Newton treated time as a universal parameter that flowed identically for all observers. He believed there was an absolute time, one that did not depend on the observer's motion or position. In practice, scientists and philosophers who followed this tradition could imagine a universal cosmic clock ticking away in the background, marking the same time for everyone. If two events occurred simultaneously in one place, it was assumed that any sufficiently informed observer anywhere would eventually agree on that simultaneity. This idea was not simply a convenience, it was deeply ingrained in the worldview of most classical physicists.

However, by the late 19th century, a series of perplexing experimental results began to cast doubt on the Newtonian picture. The most famous of these was the Michelson Morley experiment of 1887, carried out by Albert Michelson and Edward Morley. They attempted to detect the Earth's motion relative to a supposed luminiferous ether, which was then believed to be the medium through which light waves propagated. To their surprise, their carefully designed apparatus found no significant difference in the speed of light despite the Earth's orbital motion. This null result was deeply puzzling. If Newton's mechanics and Galilean relativity held true in all regimes, one might have expected to measure a variation in the speed of light depending on the direction of motion through the ether.

Over the following decades, theoretical insights by Hendrik Lorentz and Henri Poincaré, who introduced transformations to maintain the constancy of the speed of light, hinted that time and space were not as fixed as Newton imagined. By the time Einstein published his Special Theory of Relativity in 1905, the intellectual climate was primed for a radical revision.

Einstein's solution was both elegant and daring. He proposed that the speed of light in a vacuum is constant for all observers, regardless of their motion. To reconcile this absolute speed of light with experiments and logic, one had to abandon the notion of absolute simultaneity. In other words, Einstein concluded that events considered simultaneous by one observer might not be simultaneous for another observer moving at a different velocity. This suggestion was a direct blow to the Newtonian concept of a universal time.

By removing the idea of absolute simultaneity, Einstein essentially dismantled the old scaffolding that supported a single shared time applicable to all. Different observers moving relative to each other now carried their own slices of simultaneity, their own notion of what constituted the present moment. Although Einstein was still employed at the Bern Patent Office when he published these seminal papers, his revolutionary thinking reached the leading physicists of the world quickly. Many scientists found his conclusions difficult to accept at first. Some, like the mathematician and physicist

Hermann Minkowski, initially regarded Einstein's work as a curiosity. Yet Minkowski soon realized the profound geometric structure underlying Einstein's reforms and, in a 1908 lecture, introduced the modern concept of spacetime.

In just a few short years, the historical perspective had changed dramatically. Newton's universal time had crumbled, replaced by a more nuanced understanding in which time depended on one's frame of reference. The shift was as profound as any in the history of physics. The intellectual journey from absolute time to relative time served as the crucial first step toward the idea of a block universe, in which no single universal now exists.

**The Relativity of Simultaneity**

A cornerstone of Special Relativity is the relativity of simultaneity. Before Einstein, most scientists accepted without question that simultaneity was absolute. It seemed so obvious that if two events occurred at the same time according to one clock, then another identical clock, provided adequate synchronization, would mark these events simultaneously as well. Einstein, drawing upon Lorentz's transformations, showed that simultaneity is observer dependent.

To illustrate this with a famous thought experiment presented in early relativistic arguments: Imagine two lightning strikes hitting opposite ends of a moving train. An observer on the ground, standing midway between these strikes, sees the flashes at the same time and concludes they were simultaneous. An observer aboard the train, also positioned midway between the ends, sees the flashes at different times due to the motion of the train and concludes that they were not simultaneous. Both observers are correct in their respective frames of reference. This was a dramatic revelation because it removed any privileged frame or universal synchronization. Suddenly, the concept of a shared universal now became untenable.

In the early 20th century, physicists grappled with this idea, sometimes joking that time had become "elastic." Einstein himself famously remarked that if his theory caused perplexity, it was because it challenged such a natural and longstanding assumption. But experimental evidence supported Einstein's conclusions. For instance, the Ives Stilwell experiment in 1938 confirmed time dilation and indirectly reinforced the relativity of simultaneity. Later precision tests, including observations of fast moving particles in accelerators and measurements with highly accurate atomic clocks in different inertial frames, continued to confirm these insights. Over time, as these experimental verifications piled up, skepticism waned and the physics community accepted the relativity of simultaneity as a fundamental reality.

From a human perspective, letting go of absolute simultaneity can feel unsettling, yet physicists gradually embraced it as they saw the mathematical consistency and the explanatory power it provided. As one historian of science noted, no principle was abandoned by the community of physicists without a compelling reason, and the evidence pointing to relativistic effects was too solid to ignore. This was not a trivial philosophical tweak. It altered how we interpret measurements, synchronize clocks, and understand signals traveling at or near the speed of light. By the time these concepts were well established, the door was wide open to new ways of thinking about time, including the possibility that all moments might coexist in a single four dimensional structure.

## The Geometry of Spacetime

The unification of space and time into spacetime was primarily the work of Hermann Minkowski, who in 1908 presented a geometric interpretation of Einstein's relativity. Minkowski introduced a four dimensional mathematical framework, now known as Minkowski spacetime, in which time and space coordinates were treated as parts of a unified geometric entity. He famously declared that space by itself and time by itself were doomed to fade into mere shadows, but that their union would shine as a fundamental aspect of reality.

This geometric viewpoint had profound consequences. In Minkowski's spacetime, events are represented as points in a four dimensional manifold, with three spatial coordinates and one time coordinate. Instead of trying to maintain absolute time, the invariance of the speed of light and the relativity of simultaneity became simpler and more elegant when viewed through this four dimensional lens. The transformations derived by Lorentz and employed by Einstein acquired a clear geometric meaning. They could be understood as rotations and boosts in a four dimensional arena, preserving a quantity known as the Minkowski interval. This interval remains invariant for all observers, reflecting the fundamental structure of spacetime itself.

The shift from viewing space and time separately to treating them as a unified continuum was a major conceptual leap. Leading physicists and mathematicians, including Arnold Sommerfeld, Max Born, and later John Wheeler, recognized the power of Minkowski's formulation. Although early resistance to these geometric ideas existed, this approach gradually became standard in theoretical physics. The modern language of relativistic physics, whether in textbooks or research articles, is thoroughly grounded in this geometric viewpoint.

By highlighting the geometry underlying physical phenomena, Minkowski's approach offered a consistent framework for understanding relativistic effects. This geometric insight not only helped explain why simultaneity is relative, it also set the stage for new interpretations of time itself. If all events can be mapped into a four dimensional structure, then what does it mean to say that only the present is real? Minkowski's spacetime treats all points equally, with no built in arrow singling out any special moment. This paved the way for physicists and philosophers to contemplate that all of time might exist at once, like a landscape extending in every direction.

## The Speed of Light as a Cosmic Speed Limit

One of the defining features of Special Relativity is that the speed of light is constant and invariant. This fact, confirmed by numerous experiments, acts like a cosmic constraint that molds how observers measure distances and durations. The constancy of the speed of light, commonly denoted as c, implies that all inertial observers will measure light traveling at the same speed, regardless of their own motion or the motion of the source.

This rule was nonnegotiable, and so physics had to adapt. The adjustments came through phenomena like time dilation and length contraction, both rigorously derived from the Lorentz transformations. Observers in relative motion never agree on simultaneity, but their descriptions remain consistent once we accept these transformations. In other words, one observer might find that a moving clock runs slower and that moving rulers appear contracted, but another observer, moving along with those rulers and clocks, will have a perfectly ordinary experience.

Historically, the realization that c is a strict limit was resisted at first. Before Einstein, many hoped that one could chase down a beam of light and see it slow or that combined velocities would behave intuitively, adding up or subtracting as classical physics suggested. Yet every attempt to break the light speed rule has failed. Observed decays of fast moving particles, measurements on relativistic electron beams, and countless high energy physics experiments have reaffirmed the speed of light as a fundamental limit. It was this ironclad constancy that inspired Einstein's leap. By treating the speed of light as a universal invariant, Einstein forced time and space to become malleable in such a way that protected c from any tampering.

This cosmic speed limit also hints that time cannot be the universal ticking everyone once imagined. Since no information can travel faster than light, observers must abandon any notion that all events across the universe are laid out in a single universal moment. Instead, each observer organizes events differently. This realization is precisely the condition needed to start imagining that time might be a dimension akin to space and that past, present, and future may coexist in a unified structure.

### World Lines and the Idea of "All at Once"

In the relativistic framework, the history of any object in spacetime can be represented as a world line. A world line traces the trajectory of an entity through both space and time. For example, consider your own existence. If we were to record your location in three dimensional space at every moment from your birth to your current age, we could plot these points in spacetime coordinates. The result would be a continuous curve that encodes every event in your personal life history. In principle, this world line exists as a geometric object in the four dimensional continuum.

While in everyday life we experience time as a sequence of moments, the geometric view suggests that all these moments are already embedded in the structure of spacetime. Several physicists, including Hermann Weyl in the early 20th century, took these ideas seriously and argued that all events are equally real. Weyl famously wrote that the objective world simply is, and does not happen. This statement was an early philosophical interpretation resonating with the block universe concept.

Over time, various experiments and observations reinforced the value of world lines. In high energy physics, world lines are used to track particle trajectories in accelerators and bubble chambers. In astrophysics, they are essential for understanding how signals travel between distant stars and galaxies. In all these applications, the idea of spacetime as a single continuous entity is vital.

From a scientific point of view, world lines are tools that help us understand causality and structure. No observer sees the entire block all at once, but the framework of relativity suggests that there is no fundamental reason to privilege any one moment. Physicists sometimes make lighthearted comments in lectures, noting that in the block view, your future career moves or your retirement years already exist in spacetime, just not accessible to you at your current frame of reference. While such comments might be tongue in cheek, they reflect a serious point: relativity does not single out any time slice as more real than another.

## Different Observers, Different Slices

Since simultaneity is relative, different observers carve out different slices of spacetime when they define what they call the present. Each observer's definition of now is a particular cross section of the four dimensional continuum. This relativity of simultaneity implies that what one observer calls the present might include events that another observer categorizes as part of the future or the past.

Historically, this notion was met with considerable philosophical debate. Traditional views of time often relied on a universal present and a clear distinction between past and future. The idea that these divisions depend on the observer's state of motion was not easily absorbed. Physicists, however, found that these new perspectives were not only consistent with measurements, they resolved contradictions that had arisen under older assumptions.

To illustrate, after Einstein's theory gained traction, numerous tests confirmed that clocks moving at different speeds show different elapsed times. Particles produced in cosmic rays that enter Earth's atmosphere at speeds close to the speed of light live much longer than expected due to time dilation. Atomic clocks flown on airplanes record different intervals than identical clocks kept on the ground. All these findings support the idea that time is not an absolute universal parameter. Instead, time intervals and the notion of simultaneity depend on motion.

By the mid 20th century, courses in relativity had become common in graduate physics programs, and the community accepted that each observer may define their present differently. Philosophers of science, meanwhile, continued to debate the meaning of this. Some embraced interpretations that time is purely relational, others struggled with reconciling this with our psychological sense of a flowing now. Still, the key point remains: the physics is unambiguous. Different slices of the four dimensional continuum appear equally valid from the standpoint of relativity, which suggests that no preferred definition of now exists.

**Toward the Block Universe**

As historians and philosophers examined the implications of relativity, the idea of a block universe gradually emerged as one natural interpretation. Although Einstein himself often avoided deep philosophical commitments, he did make comments that hinted he leaned toward a static view of spacetime. In a condolence letter written to the widow of his longtime friend Michele Besso in 1955, Einstein remarked that for physicists who believe in relativity, the distinction between past, present, and future is only a stubbornly persistent illusion. This statement, often quoted, suggests that Einstein took seriously the notion that all points in time coexist in a single geometric structure.

Hermann Weyl also supported the eternalist perspective. Weyl argued that since relativity eliminates a universal present and treats spacetime as a four dimensional manifold, we should regard all events as equally real. Philosophers like Hilary Putnam and the physicist John A Wheeler further explored these ideas, and discussions about the block universe continue into the present century. While not all physicists or philosophers agree that the block universe is the correct interpretation, it remains one of the most coherent views consistent with Einsteinian relativity.

From a historical standpoint, the block universe concept represents a shift from seeing time as a special kind of process to seeing it as a dimension analogous to space. This idea would have been incomprehensible to Newton or to many 19th century physicists. Yet after Einstein, Minkowski, and the generations of scientists who tested their theories, the block universe interpretation stands as a serious candidate. It aligns neatly with the mathematics and the empirical facts of relativity.

Some researchers have tried to link the block universe concept with other advanced theories. In general relativity, gravity is understood as curvature of spacetime, and the block concept fits well with a geometry that does not single out any moment in time. In quantum field theory, states are defined on entire spacetime configurations rather than on a single instant, which also resonates with block universe thinking. Although open questions remain about human consciousness, free will, and how we experience time, the block universe perspective does not contradict any known physics. It simply challenges our intuitive understanding of temporal flow.

**Summary**

The journey from Newtonian physics to the relativistic block universe perspective was not a simple or a sudden shift. Historically, it began with the slow accumulation of experimental evidence that defied Newtonian predictions, culminating in the Michelson Morley experiment's null result. Theoretical developments by Lorentz and Poincaré laid the groundwork, and Einstein's Special Relativity shattered the old ideal of absolute time. The subsequent geometric formulation by Minkowski introduced the concept of spacetime, making the relativity of simultaneity and the invariance of the speed of light more transparent.

With this new framework, time lost its universal status. Different observers defined simultaneity differently, and there was no longer a single present moment for all observers. The speed of light's constancy enforced transformations that mixed space and time coordinates. Experiments like Ives Stilwell, along with countless observations in particle accelerators and atomic clock comparisons, confirmed that time and simultaneity are not absolute.

World lines offered a way to represent entire lifetimes in spacetime, and as philosophers and physicists reflected on these developments, they proposed the block universe as a natural interpretation. If all events are points in a four dimensional manifold and no preferred now exists, it becomes plausible to consider that past, present, and future exist all at once. This interpretation, known as eternalism, fits neatly with the structure of relativity, although it is not mandated by it.

The block universe is an idea that arose from careful historical and scientific inquiry. While the concept can still spark debates over philosophical nuances, it stands on a solid foundation of empirical and theoretical work. Physicists have grown comfortable with the notion that time may be just another dimension, and many see no fundamental reason to distinguish between events based on when they occur. The philosophical implications remain rich territory for debate. Some argue it challenges free will or reduces the human experience of time to a psychological construct. Others find it liberating, offering a stable view of reality that is not dependent on human perception.

In any case, relativity cleared the intellectual landscape for this viewpoint by removing the final vestiges of absolute time. The process was incremental, guided by experiments, theoretical insights, and conceptual leaps. Today, while not all scientists endorse the block universe as a metaphysical statement, few doubt that relativity makes it at least a very plausible picture. Einstein's legacy, Minkowski's geometry, and countless experimental confirmations have given us a new understanding of time. The block universe stands as one of the most striking and thought provoking interpretations arising from that understanding, posing fundamental questions about the nature of reality and our place within it.

# Chapter 3

## Reconciling the Block Universe with Quantum Mechanics

The block universe hypothesis, emerging naturally from the geometric insights of Einstein's theory of relativity, provides a framework in which all events exist together in a unified four dimensional spacetime. Under this interpretation, past, present, and future are points on a single continuum, and the concept of a flowing time is replaced by a static structure. While this approach finds strong support in the deterministic nature of special relativity, it appears to stand in stark contrast to quantum mechanics, a theory that has introduced probability, uncertainty, and observer dependent outcomes into our scientific worldview.

Quantum mechanics, developed in the early 20th century by pioneers such as Werner Heisenberg, Erwin Schrödinger, Paul Dirac, and Niels Bohr, brought forth concepts like superposition, wavefunction collapse, and entanglement. These principles challenge the classical determinism that was once taken for granted. Instead of a universe in which outcomes are fixed, quantum theory seems to offer branching possibilities and probabilistic events that come into existence upon measurement. At first glance, integrating the elegant static geometry of the block universe with the seemingly fluid and indeterminate nature of quantum mechanics might appear impossible.

However, a careful reevaluation of quantum phenomena in light of the block universe picture suggests that the tension may be more philosophical than physical. By considering quantum states and measurements as embedded in a timeless four dimensional framework, we can reinterpret probability, superposition, and collapse in ways that maintain consistency with relativity. In this effort, we draw upon both well established interpretations of quantum mechanics and more recent proposals that seek determinism beneath the probabilistic veneer. While no single interpretation has won universal acceptance, viewing quantum mechanics through the lens of the block universe has the potential to bridge conceptual divides, offering a more unified understanding of nature's underlying structure.

## The Tension Between Relativity and Quantum Mechanics

Relativity, as formulated by Einstein and further elaborated by Hermann Minkowski, treats time as one coordinate within a four dimensional geometric entity known as spacetime. In this scheme, events do not happen in a global universal now. Instead, each observer slices the block of spacetime differently, calling a particular set of events their present. Everything that can happen does happen at once within the geometry of spacetime, although different observers may disagree about which events line up simultaneously.

By contrast, quantum mechanics, forged by thinkers such as Heisenberg and Bohr, revolutionized how scientists viewed causality and determinism. Experiments by Otto Stern, Max Born's probabilistic interpretations, and eventually John Bell's inequalities underscored the fundamental role of probability and measurement. Unlike classical physics, quantum theory does not assign fixed properties to particles prior to measurement. Instead, it describes probabilities encoded in a wavefunction. Only when a measurement is made does the system appear to yield a definite outcome.

This difference seems stark. Relativity suggests a completely determined universe, one in which the entire history and future of every particle and field configuration is already embedded in spacetime. Quantum mechanics, however, demands that certain properties remain indefinite until observed. If the block universe is correct, how do we reconcile the idea that all outcomes are somehow fixed with the quantum claim that outcomes are uncertain before measurement?

Historically, many physicists struggled with this conflict. The famous Bohr Einstein debates in the 1920s and 1930s highlighted deep conceptual rifts regarding the meaning of quantum theory. While Einstein favored hidden variable interpretations or some underlying reality not captured by the standard formalism, Bohr defended the Copenhagen interpretation in which quantum mechanics is complete and probability is irreducible. As experiments in subsequent decades, culminating in tests of Bell's theorem, showed no evidence for local hidden variables, many physicists grew comfortable with the inherent randomness of quantum mechanics. Yet the fundamental tension with relativity's block universe picture remained largely unresolved.

Today, researchers consider multiple interpretations. The Many Worlds Interpretation proposed by Hugh Everett offers a deterministic, if sprawling, landscape of branching universes. Bohmian mechanics provides a deterministic picture by introducing nonlocal hidden variables and pilot waves. Objective collapse theories try to modify quantum mechanics so that wavefunctions collapse spontaneously over time. Superdeterminism posits correlations between measurement settings and hidden variables that remove the need for nondeterministic collapse. Each approach comes with implications for how we view time and reality. The block universe perspective cuts across these debates, asking how all these viewpoints would look if we start from a four dimensional, timeless ontology.

**Superposition as Timeless Possibility**

In standard quantum mechanics, a system can exist in a superposition of states. For instance, an electron can pass through both slits in a double slit experiment until it is observed or measured. Only upon measurement is a specific outcome realized. Traditionally, this is framed as a dynamic process: the system evolves according to the Schrödinger equation, and when measured, collapses into a definite state.

Within the block universe, however, we can reinterpret superposition as multiple possible outcomes existing as timeless elements of the spacetime structure. Instead of seeing superposition as an evolving uncertainty waiting to be resolved, we can treat it as a set of branches or configurations already embedded in the four dimensional continuum. Each branch corresponds to a possible outcome of the measurement. Before the observer chooses a measurement setting, these outcomes all exist in a kind of timeless tableau.

From the viewpoint of an observer traveling along a worldline, their experience of making a measurement is simply the intersection of that worldline with a particular outcome embedded in spacetime. The observer does not create the outcome at the moment of measurement. Instead, the observer encounters the outcome that was always there, fixed in the block. The appearance of superposition is a reflection of the observer's partial knowledge and the probabilistic rules governing which outcome is revealed along their specific path.

In this interpretation, probability is not about events coming into existence but about the observer's ignorance of which preexisting branch of the block they will experience as their reality.

Historically, the notion that quantum probabilities might reflect ignorance of preexisting outcomes runs counter to the Copenhagen interpretation, which treats the wavefunction as a complete description of physical reality. Yet nothing in relativity forbids us from entertaining a picture in which all outcomes exist. The many worlds approach, for example, aligns naturally with a block universe perspective. All branches of the wavefunction's evolution are present in spacetime, though observers find themselves confined to particular branches. Even deterministic hidden variable theories can fit into this framework if we consider that the hidden variables define which trajectory through the block is realized.

**Wavefunction Collapse - A Perceived Phenomenon**

Wavefunction collapse is a central mystery of quantum mechanics. According to standard theory, before a measurement the system is described by a wavefunction that assigns probabilities to various outcomes. Upon measurement, the wavefunction supposedly collapses, leaving the system in a definite state. But what exactly triggers this collapse? The Copenhagen interpretation often appeals to the measurement apparatus or the observer's consciousness as playing a role. This remains philosophically troubling. Scientists like John von Neumann, Eugene Wigner, and later Henry Stapp grappled with the question of where to draw the line between quantum system and observer.

From the block universe perspective, collapse does not need to be a real dynamical event. Instead, what we call collapse might be a subjective perspective arising from our limited viewpoint as beings who experience time sequentially. In a timeless spacetime, all outcomes of a quantum experiment are fixed. What we label as collapse may just be the observer's realization of which outcome was predetermined along their worldline. There is no universal collapse event. There are only worldlines intersecting particular outcomes that were always part of the four dimensional structure.

This interpretation does not deny the predictive successes of quantum mechanics. The probabilities calculated by the wavefunction remain correct. However, rather than treating collapse as a physical process that alters reality, the block universe perspective treats it as an epistemic update for the observer. Before the measurement, the observer does not know which branch they will encounter. After the measurement, they do. The wavefunction then becomes more like a mathematical tool for calculating probabilities of which preexisting outcome will be found, not a complete dynamical state evolving in time that must collapse.

Consider entanglement, which Einstein, Podolsky, and Rosen used in their 1935 thought experiment to question the completeness of quantum mechanics. Two entangled particles exhibit correlations that seem instantaneous and not limited by the speed of light. In the block universe, these correlations are not transmitted signals but are simply the geometric relationship between entangled worldlines and events. The outcomes on both sides of the experiment were always determined, encoded in the structure of spacetime. The illusion of collapse traveling instantaneously across space vanishes if we view the entire set of outcomes as coexisting timelessly. The constraints imposed by Bell's theorem still apply, telling us that no local hidden variable theory can replicate quantum predictions. But these constraints do not prevent a block universe interpretation, where nonlocality is not a signal but a preexisting pattern in the four dimensional tapestry.

## Bridging Relativity and Quantum Mechanics

The reconciliation of relativity and quantum mechanics is one of the great open problems in modern physics. While the block universe addresses the puzzle of time, it does not by itself provide a complete unification of general relativity and quantum field theory, the framework physicists use to describe elementary particles. Still, the block universe viewpoint offers a conceptual platform that can help alleviate some of the philosophical tensions.

Deterministic interpretations of quantum mechanics, like Bohmian mechanics, find a natural home in the block universe. In Bohm's theory, particles follow well defined trajectories guided by a pilot wave. All these trajectories can be laid out in spacetime, giving a deterministic structure that fits well with a block universe. Even the Many Worlds Interpretation, which posits that every quantum outcome spawns a new branch of the wavefunction, can be visualized as a complex four dimensional structure, with countless branches coexisting. Observers find themselves on particular branches, and their experience of collapse is replaced by a realization that they occupy just one branch among many. This picture is already eternalist and dovetails with a block interpretation.

For interpretations that maintain genuine randomness, like the Copenhagen interpretation or objective collapse theories, the block universe idea is more challenging, but not impossible. One might imagine that what we call randomness is simply our lack of access to the underlying deterministic structure or that some cosmological initial conditions fix the entire block, making probabilities reflect a pattern that looks random at the local level but is fixed globally. Some researchers explore superdeterminism, which posits that measurement settings and particle properties are correlated in a way that removes the need for nonlocal influences while preserving the appearance of randomness. In a block universe, superdeterminism could be seen as simply part of the overall spacetime structure, with no conflict in principle.

As physicists continue to search for a quantum theory of gravity, often considered the holy grail of theoretical physics, the block universe perspective may gain further significance. Most approaches to quantum gravity, from string theory to loop quantum gravity, attempt to unify space and time at a fundamental level. If successful, they may produce a framework even more compatible with a block universe view. At that point, the distinction between quantum events and classical spacetime geometry might blur, and the block universe could emerge as a natural setting for a fully unified theory.

**Implications for Free Will and Existence**

A common worry is that the block universe renders all events fixed and predetermined, leaving no room for free will. When we add quantum mechanics to the mix, the tension can become even more acute. On one hand, quantum theory seems to allow genuine unpredictability. On the other, a block universe suggests that every future event, including our decisions, already exists in spacetime.

One approach to this dilemma is to distinguish between fundamental determinism and emergent phenomena. Much like temperature or pressure, free will could be an emergent concept that arises from the complexity of human brains and decision making processes. Even if all events are embedded in spacetime, the experience of making choices and holding intentions remains meaningful at the human scale. The block universe does not prevent us from experiencing a sense of agency, any more than knowing that a recorded movie has a fixed sequence of scenes prevents us from engaging emotionally with the story as it unfolds.

From the perspective of an observer moving through the block, each decision point is encountered without knowledge of the future events encoded ahead. This lack of foreknowledge ensures that subjective experiences of choice and uncertainty remain intact. Quantum mechanics, with its probabilistic predictions, can reinforce this feeling of openness. While the block may contain all outcomes, we do not have immediate access to them. The complexity of the underlying structure is beyond our capacity to fully comprehend, allowing room for the subjective impression of free will and moral responsibility.

Moreover, viewing quantum phenomena as aspects of a timeless spacetime structure invites a broader reconsideration of what existence means. Instead of seeing reality as a three dimensional world that changes over time, we consider a four dimensional entity that simply is. Our lives are worldlines, stretching from birth to death, and all events of our personal histories exist at once. The inclusion of quantum mechanics in this framework adds richness and subtlety. Instead of a dull, rigid block, we have a block filled with intricate patterns of probabilities, correlations, and structures that define what outcomes are found at which spacetime points.

## A Unified Perspective

Relativity and quantum mechanics are often presented as theoretical frameworks that clash, one providing deterministic geometry and the other probabilistic outcomes. Yet the block universe suggests a perspective in which the two can be viewed as complementary aspects of a single reality. In this picture, relativity gives us the scaffolding: a four dimensional continuum in which everything exists. Quantum mechanics fills that continuum with intricacies, granting it patterns of possible outcomes, correlations, and entanglements.

This integration does not solve every problem. It does not immediately yield a full quantum theory of gravity, nor does it end the interpretational debates in quantum mechanics. However, it does offer a way to think about quantum mysteries that is consistent with the relativistic view of spacetime. The block universe reframes quantum paradoxes. Instead of worrying about how or when the wavefunction collapses in time, we ask how events are arranged in spacetime. Instead of struggling with faster than light influences, we recognize that entanglement correlations are geometric facts about the block, not signals traveling through it.

Historical progress in physics often comes from adopting new viewpoints. Einstein's special relativity emerged from the decision to hold the speed of light constant and accept that time intervals could vary for different observers. General relativity arose when Einstein recast gravity as a manifestation of curved spacetime rather than a traditional force. Quantum theory developed when scientists accepted that particles could have probabilistic descriptions and that measurement plays a fundamental role. The block universe interpretation might not be the last word, but it continues in that tradition of conceptual innovation, offering a viewpoint that might help us better understand the deep structure of reality.

The human aspect of this search should not be overlooked. Physicists are people too, with intuitions shaped by everyday experiences. The notion of a timeless four dimensional structure where quantum uncertainties are just patterns of outcomes can feel abstract. It can sometimes lead to a sense of philosophical unease, as if we are losing the comforting narrative of time passing. Yet the reward is a more coherent understanding of how two of our best tested frameworks fit together. If this approach helps us imagine that behind the scenes, reality is a richly woven tapestry, then that imaginative leap may guide future discoveries.

In adopting this perspective, we see time not as a flowing river of change, but as a landscape where every point is fixed in place. Within that landscape, quantum mechanics describes the distribution of possible outcomes, while relativity fixes their geometric relationships. Taken together, they form a vision of reality that is at once deterministic in structure yet filled with intricate patterns of probability and correlation.

This might not settle every question about interpretation, nor close the book on philosophical puzzles. But it does encourage us to think more freely, to consider that what appears as randomness in our local, time bound experience is part of a larger determinism embedded in the structure of the universe. Rather than clashing perspectives, relativity and quantum mechanics may be two complementary lenses that, when combined, give us a richer, more complete view of the underlying reality.

In this sense, the block universe hypothesis, reconciled with quantum mechanics, can be seen as a conceptual tool. It invites us to embrace the complexity of nature, rather than trying to force it into familiar categories of deterministic or probabilistic, timeless or evolving. If we accept that all events exist in the four dimensional fabric of spacetime, then quantum phenomena become features of this fabric, giving it texture and depth. The universe, in this view, is an elegant and interconnected whole, a complete structure that includes every event and every observation, linked together in a grand design beyond our limited temporal standpoint.

# Chapter 4

## Philosophical and Metaphysical Underpinnings of Eternalism

Eternalism, as a philosophical and metaphysical standpoint closely aligned with the block universe interpretation, offers a fundamental challenge to our usual notions of time. Traditionally, we think of the present as the only truly existing moment, with the past irretrievably lost and the future not yet real. Eternalism, however, insists that all points in time, whether past, present, or future, have equal ontological status. In other words, every historical event and every future possibility coexist with current happenings in a vast four dimensional spacetime structure. This perspective finds strong resonance with modern physics, particularly Einstein's theory of relativity, which undermines the notion of a universal now and suggests that temporal divisions depend on the observer's frame of reference.

Despite eternalism's scientific appeal, the concept can feel intellectually counterintuitive and emotionally unsettling. Does a world in which all times exist at once diminish the importance of the present moment? How does eternalism affect our understanding of free will, ethics, and personal identity? Could this static view of time fit with human experience, morality, and even spirituality? These questions have engaged philosophers, theologians, and scientists for decades. While some thinkers embrace eternalism's elegance and compatibility with contemporary physics, others resist it, arguing that it neglects the undeniable sense of time's passage in human life.

In this chapter, we will explore the philosophical and metaphysical foundations of eternalism, detailing how it contrasts with alternative views like presentism and the growing block theory, and why some philosophers and scientists consider eternalism a natural outgrowth of our best physical theories. We will also examine the implications of eternalism for our understanding of becoming, free will, memory, morality, spiritual perspectives, and the broader meaning of existence.

**Presentism, Growing Block, and Eternalism: Three Views of Time**

To appreciate eternalism fully, it helps to locate it among the three major philosophical theories of time: presentism, the growing block theory, and eternalism itself.

**Presentism**: Presentism claims that only the present moment is real. The past is gone, existing now only as memories or records, and the future has not yet emerged into existence. This view matches everyday intuition. We feel time "flows" from past to future, and at any given moment, only now is concretely real. Historically, presentism aligns well with pre relativistic physics, where time was absolute and universal, and it resonates with ordinary human language, which treats the past as something that used to be but no longer is.

**Growing Block Theory**: The growing block theory tries to balance the realism of past events with the openness of the future. According to this viewpoint, the past and the present exist, but the future does not. Time is like a block of existence continually extending forward as new events come into being. Imagine a growing tower of wooden blocks, where each new block added on top represents the current moment. What lies below exists, what lies at the top is now, and above that, there is empty space for the as yet nonexistent future. The growing block model attempts to preserve an objective sense of becoming and evolution without relinquishing the reality of past events.

**Eternalism**: Eternalism posits that all points in time, past, present, and future, are equally real. Rather than a spotlight illuminating only the present or a tower that grows over time, eternalism envisions a completed four dimensional structure. If presentism is like a single painting lit by a lamp and the growing block is like a structure under construction, eternalism is like a fully constructed museum containing all paintings at once. Observers only experience certain artworks at any given moment due to their location within the structure, not because the unvisited galleries do not exist.

Eternalism is closely connected to the block universe interpretation in physics. Special relativity's relativity of simultaneity suggests there is no universal now. Different observers moving at different velocities disagree on which events are simultaneous, making it hard to maintain a unique global present. Eternalism capitalizes on this, arguing that since physics treats all events as points in a four dimensional spacetime, metaphysics should follow suit and accept that all times are equally real.

**Metaphysical Foundations of Eternalism**

The metaphysical rationale behind eternalism often begins with a straightforward question: why suppose only the present exists if fundamental physics treats time much like space, as one dimension of a larger continuum? In Newtonian mechanics, time was absolute and universal, making it plausible to conceive of a single evolving present. But relativity undermines that foundation, removing any absolute notion of simultaneity. If there is no single preferred slicing of spacetime into moments, then how can we cling to the idea that only one moment is real?

Philosophers like Hilary Putnam and W.V.O. Quine, engaged in debates during the mid 20th century, argued that because our best theories of physics present time as a dimension, we should take that dimension seriously. Putnam, for example, used arguments based on relativity to show that since what one observer calls "now" includes events another observer calls "past" or "future," the distinction between past, present, and future cannot be fundamental. From a metaphysical standpoint, it becomes natural to treat all events as existing. Quine's ontological minimalism and commitment to physicalism also support a four dimensional ontology, where no special metaphysical privilege is granted to the present.

In more recent years, philosophers of science have continued to refine these arguments. The reasoning is not just about theoretical elegance; it is about aligning metaphysics with empirical evidence. Since the laws of physics do not single out a moving now or a global present, metaphysics should not add such a structure without necessity. Eternalism thus emerges as the simplest ontology consistent with relativity. It does not require positing any flow of time or a special mechanism of becoming; it treats all moments as coexisting facts.

### The Illusion of Temporal Flow

A major stumbling block for accepting eternalism is our intuitive conviction that time flows. We feel time passes, that we move from past to future, and that the present is unique. Eternalists respond by framing this as a psychological phenomenon rather than a physical one. They argue that our experience of time's passage arises from the structure of human cognition, memory, and perception, not from an objective temporal flow in the world.

Historically, philosophers as early as Parmenides in ancient Greece challenged the idea of temporal becoming. In the early 20th century, thinkers like John McTaggart famously labeled the notion of a flowing time as an illusion. McTaggart distinguished between different ways of describing time, the A-series (past, present, future) and the B-series (earlier, later), and concluded that the flow we attribute to the A-series was not coherent. Eternalists often invoke McTaggart's reasoning to support the claim that temporal flow is a subjective construction.

From a cognitive perspective, consider how the brain encodes memories. We remember past events, carry records and evidence of them, but we have no memory of future events. This asymmetry creates an impression of directedness. Many scientists link this impression to thermodynamics: entropy increases as we move from what we call the past toward the future, giving us an arrow of time at the macroscopic level. Our psychological impression of time's flow emerges from this thermodynamic arrow and from the fact that we accumulate memories of earlier states, not later ones. Eternalism suggests that while we perceive a flow, no such objective flow exists in the four dimensional structure.

## The Problem of Free Will

One of the most contentious philosophical challenges to eternalism concerns free will. If the future already exists, can we still claim to have genuine choice? The idea that future events are as real as present events might seem to imply a fixed destiny. From a human standpoint, this is troubling because it suggests that all our deliberations and decisions are inscribed in spacetime, leaving no room for alternative possibilities.

However, eternalists often note that determinism and eternalism are distinct concepts. Determinism is about whether the laws of nature, along with initial conditions, fix a unique outcome for every future event. Eternalism is about what exists, not necessarily about how it is caused. It is conceivable to have a block universe that includes indeterministic quantum events, branching possibilities, or complex structures that reflect genuine randomness. Even in a deterministic setting, the tension between free will and a fixed future predates eternalism; it troubled philosophers and theologians for centuries in classical mechanics and theological discussions of divine foreknowledge.

Some eternalists argue that free will resides in the agent's capacity to respond to reasons and internal states. The fact that the future is "there" does not mean we cannot behave freely. Instead, our free decisions, from an eternalist viewpoint, are part of the block. In other words, our choices, though fixed in the block, are not forced upon us in any psychologically salient sense. They reflect who we are as agents. Just as the shape of a landscape exists, but hikers still choose their paths through it, our decisions define which temporal slice we find ourselves contemplating. The existence of a future slice does not undermine the authenticity of the decision making process.

Philosophically, this debate is far from settled. Some thinkers remain uncomfortable with any notion that the future is as real as the present. They worry this empties moral responsibility of content. Others find comfort in the idea that even if the future exists, we do not know it yet, and we still must deliberate and act based on our beliefs and desires. Jokes occasionally arise in philosophy seminars: "If all my finals are already graded somewhere in the block, does it matter if I study?" The eternalist answer is that your studying is also part of the block, and your effort directly influences which future slice you, as a conscious observer, will experience as your present. This humor, while lighthearted, underscores that timeless existence does not necessarily negate the meaningfulness of action.

## The Nature of Change and Becoming

Another philosophical challenge is how eternalism accounts for change and becoming. After all, change is central to human experience. We witness growth, development, and transformation. A seed becomes a plant; a child becomes an adult. If all moments exist at once, where is the becoming?

According to eternalism, change is about comparing different slices of the four dimensional entity. Just as different pages of a novel reveal different states of the story's characters, different temporal slices of the block reveal different configurations of matter and energy. The block does not change; it simply is. Our experience of becoming is tied to our perspective as beings embedded in one slice at a time. In other words, the sense of change arises because we only perceive one temporal cross section of the block at once, and we remember having encountered earlier slices that differ from the current one.

Historically, philosophers have given careful thought to this issue. In the 20th century, the philosopher Adolf Grünbaum and later thinkers considered how relativity and eternalism reshaped our notions of temporal becoming. Grünbaum argued that the objective becoming concept is incompatible with relativity, reinforcing eternalism's stance. If becoming were objective, it would single out a unique set of simultaneities that would correspond to what "just came into being," but relativity denies any such universal simultaneity. Hence, if we trust physics, a global becoming process cannot be fundamental.

Eternalists thus see becoming as an emergent phenomenon related to the internal states of observers who carry memory traces of previous slices. The narrative of a changing world is an internal construct rather than an external fact about time's flow.

**Memory, Experience, and the Psychological Arrow of Time**

If eternalism holds that all moments exist equally, why do we only recall the past and not the future? This asymmetry is crucial to our psychological sense of time. Eternalists point to physical laws that are largely time symmetric at the fundamental level, yet at the macroscopic scale we see irreversible processes driven by thermodynamics. Entropy, a measure of disorder, almost always increases as we move from what we label past to future. This entropy gradient leaves records and traces that encode information about earlier states. We carry these records in our brains, forming memories of past experiences.

This creates what philosopher Huw Price and other contemporary philosophers of science describe as the psychological arrow of time. Our knowledge accumulates from past slices, not future ones. Eternalism does not deny this arrow; it explains it as a result of physical and cognitive structures. The block contains both directions in time, but the distribution of low entropy states early in the universe sets a direction in which memory is possible. Thus, we remember what we call the past because that is where the low entropy conditions allowed stable records to form. The future, equally real in the block, remains unknown to us, not because it does not exist, but because it is not encoded in our current memory structures.

From a personal perspective, this can feel strange. I, for instance, clearly recall details from my childhood, summers spent helping my grandmother tend her garden, learning how to ride a bicycle, yet I have no corresponding memories of my old age (assuming I reach it). Eternalism would say that old age slice is there, as a part of the block, but my current brain state does not have the encoded information from that future slice. My memories reflect the order in which I encounter slices of spacetime, preserving the illusion that time flows forward. My personal anecdote illustrates how eternalism can coexist with everyday experiences. I still cherish my past and remain curious about my future, even if both are eternally set in the block structure of reality.

**Eternalism and the Nature of Reality**

Eternalism encourages a profound shift in how we conceive reality. Instead of imagining a three dimensional world that changes through time, eternalism suggests that reality is fundamentally a four dimensional spacetime manifold, static and complete. Our intuitive notions of transience, passing moments, and temporal depth give way to an ontological symmetry between past, present, and future.

In the early 20th century, philosophers and scientists considered the implications of Einstein's relativity on metaphysics. Hermann Minkowski's geometric formulation of spacetime inspired many thinkers to treat time as a dimension akin to space. Just as the entire spatial expanse of the Earth exists even though we occupy only one location at a time, so too the entire temporal extent of the universe may exist, though we perceive only one moment at a time. Quine and Putnam, among others, saw no reason to privilege a temporal slice over another. If a spatially distant mountain exists even though we are not currently looking at it, why not grant the same status to a temporally distant event?

From a philosophical standpoint, eternalism aligns with a kind of four dimensional realism known as the "B-theory" of time. The B-theory holds that all times are equally real and that temporal order is a matter of relations like earlier than and later than, not an absolute coming into existence. Under this theory, the universe is a spacetime landscape we survey slice by slice. Although this may not conform to our daily language and emotional intuitions, advocates argue that it makes better sense of modern physics and avoids unnecessary metaphysical complications.

## Eternalism's Rivals and Their Critiques

Not all philosophers and physicists are convinced by eternalism. Presentists defend the intuitive idea that only the present is real. They argue that eternalism neglects the felt reality of becoming and fails to capture the undeniable fact that we live in a world where new events seem to arise continuously. To presentists, eternalism's insistence that all moments exist at once seems to flatten the richness of temporal experience.

The growing block theory, meanwhile, tries to preserve some aspects of our intuition. It acknowledges that the past and present exist but denies reality to the future. Proponents believe this allows for a sense of genuine becoming, as the block extends forward in time. Critics of the growing block theory, however, point out its arbitrariness. Why stop at the present? Relativity's lessons apply equally to future events, undermining the notion that there is a special boundary where reality stops.

Another common critique is that eternalism makes time too much like space. Time differs from space because of its directional features and irreversibility. Eternalists respond that these features are emergent rather than fundamental. While space and time differ in some respects (such as the metric signature in relativity), the fundamental geometry treats them both as dimensions of a unified manifold. The arrow of time, they argue, emerges from conditions related to entropy and boundary conditions in the universe's initial state, not from any metaphysical flow.

Philosophers such as Tim Maudlin have argued that the block universe picture is too passive, failing to capture the dynamic aspects of physical laws. He contends that laws express how states evolve over time, and treating time as just another dimension might neglect the explanatory role of time in physics. Eternalists counter that even the laws can be viewed as relations between different slices of spacetime. The debate is ongoing and underscores that the metaphysics of time is far from settled.

## Ethics and Significance in an Eternal World

If the future is fixed, does this undermine our moral intuitions or sense of purpose? On the face of it, knowing that all events are laid out might seem to strip our actions of significance. Yet eternalists often argue that significance and responsibility remain intact. The future's existence in the block does not negate the fact that our current actions are part of the block's structure. If I choose to donate to charity today, that choice and its consequences exist in the block, shaping those future slices where the charity's beneficiaries enjoy improved conditions.

Ethically, one could say that eternalism changes our perspective. Instead of seeing morality as a tool to shape a not yet existing future, we might view it as our participation in a four dimensional tapestry. We have a role to play, and our behavior is woven into the fabric of reality. This can be uplifting: just as the chapters of a well-crafted novel contribute to a cohesive narrative, our moral choices add to the overall structure of the block. Philosophers like D. H. Mellor have discussed how B-theory perspectives, similar to eternalism, do not remove moral or personal relevance, since subjective experience and moral deliberation are local phenomena that still matter to the agents involved.

Even humorous perspectives can help: "If my future self is already enjoying the fruits of my current hard work, can I sit back and do nothing?" Eternalists respond that your hard work is part of why that future slice is positive. The block includes your diligent actions and their outcomes. You remain accountable for your choices, as they define the version of the future included in the block. Much like knowing a story is already written does not remove the author's responsibility in having created it, knowing the future is fixed in the block does not remove the moral weight of your decisions in shaping it.

**Spiritual Reconciliation**

Eternalism's implications also resonate with certain spiritual and theological traditions that consider a timeless or eternal perspective on reality. Some religious views hold that a divine entity sees all events, past and future, simultaneously. In Christian theology, the idea of God's foreknowledge often suggests that God perceives all of history at once. Eternalism offers a naturalistic parallel to this idea, suggesting a universe in which all moments are laid out, and a divine perspective could encompass the totality without the constraints we face as temporal observers.

This notion can be comforting: it implies that nothing is ever truly lost to time, since all events remain part of the four dimensional structure. From a Christian perspective, for example, the narrative of Jesus Christ's life, death, and resurrection, as mentioned in the original text, can be viewed as eternally present within the block. The divine plan, moral teachings, and spiritual guidance offered by religious figures may exist timelessly. Believers can interpret eternalism as consistent with the idea that God is outside of time and that the human experience of temporal flow is a limited perspective on a grander, eternal reality.

Such reconciliation is not universal. Some religious traditions emphasize the importance of becoming, growth, and change as essential features of moral and spiritual development. They might find eternalism's static ontology less appealing. Others worry that eternalism could diminish the role of prayer, hope, and repentance if the future is as fixed as the past. Yet others see it as a way to affirm that all moments of moral striving and spiritual insight remain real and meaningful, underscoring the eternal value of virtuous acts.

The dialogue between eternalism and spirituality is ongoing. While eternalism is not inherently religious, it shares structural similarities with certain theological notions of eternity. Philosophers and theologians can find fertile ground in these discussions, whether they seek to integrate eternalism into their religious worldview or oppose it on theological grounds. The interplay between eternal metaphysics and spiritual conceptions of time invites a richer understanding of how people interpret existence at its deepest levels.

## Eternalism's Broader Impact and Ongoing Debates

As we have seen, eternalism encourages rethinking some of our most basic intuitions about time. It aligns smoothly with the block universe picture suggested by relativity, offers a coherent explanation for why the present is not universally privileged, and provides a conceptual framework that accommodates the psychological impression of time's flow, free will, ethics, and even spiritual perspectives.

However, eternalism does not settle all issues. The nature of the block universe and how exactly it reconciles with quantum mechanics, as discussed in previous chapters, remains an active area of inquiry. Philosophers continue to debate whether eternalism fully captures the complexity of human experience. Scientists probe whether the universe might have more subtle structures, like causal sets or emergent temporal directions that challenge a simplistic block view.

From a personal standpoint, one might ask what it means to live in a universe where one's entire life, from birth to death, is laid out in the spacetime continuum. Some find comfort in knowing that cherished moments are never truly lost. Others find it disconcerting, feeling it robs life of spontaneity. On a lighter note, some might joke that eternalism means we can blame our bad decisions on the fact that they are "just there" in the block. Of course, eternalists would retort that we are still morally responsible, since our decisions form part of the structure that defines us as agents.

In academic philosophy, debates between eternalists, presentists, and advocates of the growing block theory remain lively. Each camp publishes articles, responds to criticisms, and refines its arguments. At conferences and in journals, these philosophers examine relativity's implications, the logic of temporal predication, the metaphysics of causality, and the phenomenology of temporal experience. The conversation shows no sign of abating. Indeed, as physics evolves, if we discover new principles of quantum gravity or cosmological structure, it might strengthen or challenge eternalism further.

## Summary

Eternalism is a robust metaphysical and philosophical position that invites us to reconsider the nature of time, existence, and reality. By claiming that past, present, and future are equally real, eternalism harmonizes with the block universe model implied by special relativity, giving a metaphysical interpretation that does not rely on a universal now or a flowing temporal dimension. Instead, all events lie frozen in a four dimensional spacetime landscape.

This perspective reshapes our understanding of fundamental concepts. Becoming, change, and the flow of time emerge as subjective experiences rather than objective features of reality.

Free will, rather than requiring an open future, can be reinterpreted as the capacity to choose paths embedded within the block, making our decisions no less meaningful. Morality and ethics remain crucial, as our actions define the structure of the block. Memory and the arrow of time find explanation in thermodynamic gradients, not in a metaphysical property of the universe. Spiritually, eternalism resonates with timeless theological visions, offering a conceptual parallel to the idea of a divine vantage point outside of time.

This is not to say eternalism is without controversy. Presentists and growing block theorists contest its dismissal of temporal passage. Critics argue that eternalism cannot adequately capture the dynamic richness and existential texture of human experience. Others worry that making time too space like oversimplifies the distinctive features of temporal direction, causality, and irreversibility.

Nevertheless, eternalism provides an elegant, scientifically informed metaphysical foundation from which to explore deeper questions about the universe. By embracing a four dimensional ontology, we gain a framework for asking whether we can navigate the block universe, manipulate aspects of time, or perhaps reconcile quantum mechanics and gravity in a unified picture. Although these remain speculative frontiers, eternalism ensures that when we contemplate them, we do so from a vantage point consistent with the best physics of our era.

As we continue on this intellectual journey, eternalism invites us to see ourselves not as travelers swept along by a temporal current, but as inhabitants of a vast and complex spacetime. We are creatures embedded in a four dimensional tapestry, experiencing events slice by slice. Whether we find this comforting, disconcerting, or exhilarating, eternalism ensures that our inquiries into time's nature remain anchored in a coherent metaphysical vision aligned with the scientific image of the world.

# Chapter 5

## Observational Evidence and Theoretical Support

When we contemplate the block universe, we face a conceptual challenge unlike the direct spatial vistas we enjoy in everyday life. While standing on a hillside we can survey distant hills and rivers, confident that what we see spatially exists, we have no comparable visual ability to examine the future or past directly. We rely instead on scientific evidence, theoretical consistency, and experimental results to guide our understanding of time. In a world that often feels defined by transience and motion, the block universe hypothesis asserts that all events, past, present, and future, have an equal claim to existence in a four dimensional continuum.

In previous chapters, we discussed relativity, quantum considerations, and philosophical arguments that support eternalism. Now, we turn to a more empirical and theoretical analysis. How do observational data, precise experiments, and well tested physical theories bolster the block universe perspective? Much of the evidence we will survey comes from relativity, both special and general, and from practical experiments that confirm relativistic predictions. These findings strongly support the four dimensional geometric picture of spacetime, dissolving the notion of a privileged present and affirming that time should be treated as a dimension akin to space.

While we lack a "time telescope" that lets us directly view future events in the same sense we see distant galaxies, we do have many indirect lines of evidence. The constancy of the speed of light, time dilation experiments, the relativity of simultaneity, and the curvature of spacetime as revealed by general relativity all point to a unified four dimensional structure. Observations in astrophysics, the success of technologies like GPS, and the geometric nature of quantum field theory further underscore that time is woven into the fabric of reality. Taken together, these results build a compelling empirical and theoretical case for the block universe.

**The Speed of Light as a Cosmic Yardstick**

One of the earliest and most influential pieces of evidence supporting a block universe framework comes from the constancy of the speed of light. Historically, physicists in the late 19th century sought to measure differences in the speed of light relative to an assumed "ether," a hypothetical medium through which light waves were thought to propagate. The famous Michelson Morley experiment of 1887 attempted to detect variations in light speed depending on Earth's motion through the ether. Unexpectedly, the experiment failed to find any difference, strongly suggesting that the speed of light was invariant regardless of the observer's motion.

This result was deeply puzzling. Newtonian mechanics suggested that if you moved toward a light source, you should measure a greater speed of light, just as moving toward a sound source causes the sound's frequency to increase. Yet nature refused to comply. Einstein's solution in 1905, with the introduction of special relativity, was to jettison the notion of absolute space and time. Instead, he assumed the invariance of the speed of light as a fundamental principle. From this principle, Einstein derived the Lorentz transformations which show how space and time coordinates of events change between observers moving relative to one another.

Why does this support a block universe perspective? Because once you accept that the speed of light is constant and treat it as a fundamental invariant, you are led to a geometric interpretation of spacetime. The relativistic transformations mix space and time coordinates in a way that is mathematically analogous to rotations in a four dimensional space, known as Minkowski spacetime. This geometric picture, articulated clearly by Hermann Minkowski in 1908, shows that events form a four dimensional continuum. In such a geometry, all points in time exist together in the same sense that all points in space coexist. There is no absolute division of events into past, present, and future that all observers would agree upon. Instead, time must be treated as a dimension fully integrated with space.

**Time Dilation and Length Contraction: Experimental Realities**

In addition to the constancy of the speed of light, special relativity predicts time dilation and length contraction. These effects have been tested repeatedly. One of the earliest convincing pieces of evidence came from observations of cosmic ray muons in the mid 20th century. Muons are subatomic particles created by collisions of cosmic rays with molecules in Earth's upper atmosphere. They have a short lifespan of about two microseconds in their rest frame. If we treated time classically, most muons would decay long before reaching Earth's surface. Yet, we detect muons at sea level in large quantities. Why? Because from Earth's perspective, the muons' internal clocks run slower due to their high speed. In other words, time dilation allows them to survive longer and reach the ground.

Particle accelerators in modern physics laboratories provide further, highly controlled tests. When we accelerate particles like electrons or protons to near the speed of light, their internal processes slow down relative to an observer at rest. Their effective lifetime increases exactly as predicted by special relativity. The agreement between theoretical predictions and experimental outcomes is remarkable, confirming that time depends on motion.

These experiments show that time is not an absolute universal parameter. If different observers measure different elapsed times for the same process, the notion of a single, objective present for the entire universe becomes untenable. Instead, each observer follows a unique worldline through spacetime, and what they consider simultaneous events may differ. This shifting relationship between space and time coordinates aligns naturally with the block universe view, in which all events exist in a unified four dimensional framework. The variation in measured time intervals is a natural consequence of slicing this four dimensional structure differently depending on an observer's state of motion.

**The Absence of a Universal Now**

Another powerful piece of evidence supporting the block universe idea is the relativity of simultaneity. Experiments and observations have confirmed that whether two spatially separated events occur at the same time depends on the observer's frame of reference. This is not just a theoretical curiosity. It is deeply embedded in the equations of special relativity and has been tested through experiments measuring the synchronization of clocks and timing signals.

For example, consider two distant lightning strikes. One observer, standing equidistant from the strikes, may judge them to be simultaneous. Another observer, moving rapidly relative to the first, may find that one strike occurred before the other. Both observers' measurements are equally valid, and there is no absolute fact of the matter about which event occurred first. This is a radical departure from Newtonian mechanics, where simultaneity was absolute.

The relativity of simultaneity is crucial for the block universe interpretation. If there were a universal now slicing the entire cosmos at once, all observers would agree on simultaneity. The fact that they do not strongly suggests that no such global present exists. Instead, each observer's definition of simultaneity corresponds to a particular slice of the four dimensional block of spacetime. This explains why eternalism and the block universe view are so appealing: they dispense with the need for a universal present and treat time as another dimension, just like the three spatial ones.

**General Relativity and the Curvature of Spacetime**

Einstein's General Theory of Relativity, formulated in 1915, further cements the four dimensional geometric viewpoint. General relativity describes gravity not as a force acting at a distance, but as a manifestation of curved spacetime. Mass and energy deform the geometry of spacetime, and free falling objects follow geodesics in this curved structure. Observational tests of general relativity began early, with Arthur Eddington's 1919 expedition measuring starlight bending around the Sun during a solar eclipse. The agreement with Einstein's predictions was a triumph, convincing the scientific community of general relativity's validity.

Subsequent confirmations include the precise perihelion shift of Mercury's orbit, gravitational redshift measurements, the Pound Rebka experiment verifying gravitational time dilation, and most spectacularly, the detection of gravitational waves by LIGO in 2015. All these phenomena are understood naturally in a spacetime framework where time is not separate and absolute but part of a dynamic, curved geometric entity.

From a block universe perspective, general relativity extends the geometric interpretation of special relativity. Instead of a flat Minkowski spacetime, we have a curved manifold that can be warped and stretched. Yet the underlying idea remains: time is woven into the geometry. In a block universe interpretation, all points in spacetime exist as part of the four dimensional structure. The curvature of spacetime affects the measurement of intervals and the paths of particles, but it does not single out a present moment. The entire spacetime manifold, including what we call past and future, is there in the geometry.

**Cosmic Observations and the Expanding Universe**

Astronomy and cosmology provide additional, though more indirect, support for the block universe idea. When we observe distant galaxies, we are looking into the past. The light we see today left those galaxies millions or billions of years ago. The cosmic microwave background (CMB), discovered in 1965 by Arno Penzias and Robert Wilson, is a snapshot of the universe when it was only about 380,000 years old. This thermal radiation reaches us now, giving us a window into the distant past.

If time were a universal flow, one might be tempted to imagine all observers sharing a common universal moment. Yet the fact that we see different snapshots of the universe at different epochs depending on the direction and distance we look suggests that the division into past, present, and future depends on our vantage point. Distant events that we label as past because their light is only now reaching us are just as much part of the four dimensional structure as local events. The universe does not provide a universal stage with a single time coordinate that everyone agrees upon; rather, it offers a web of events, each with its own spatio-temporal coordinates.

The expanding universe model, supported by Edwin Hubble's 1929 discovery of galactic recession and refined by modern cosmology, also fits well with eternalism. Instead of an evolution of the cosmos from a past Big Bang toward a future expansion as a single unfolding narrative, the block universe perspective treats all epochs, Big Bang, earlier eras of galaxy formation, current age, and distant future states, as parts of a four dimensional structure. While we cannot move freely through time as we move through space, the geometry remains consistent with a timeless view in which all moments co-exist in the block.

**Quantum Considerations**

Quantum mechanics introduces probabilities and uncertainties that might seem at odds with a static block. Historically, many physicists, including Niels Bohr and Werner Heisenberg, emphasized the role of measurement and the collapse of the wavefunction. This appears to suggest that the future is not fixed and that the universe is in some sense "incomplete" until observed.

However, interpretations of quantum mechanics vary widely. Many worlds theory, proposed by Hugh Everett in 1957, removes the notion of collapse and treats all outcomes as equally real, branching into parallel worlds. This picture, while speculative, aligns surprisingly well with an eternalist view. In a four dimensional block, all possible outcomes could exist as part of a grander structure, possibly extending into a higher dimensional manifold. Even without committing to many worlds, relativistic quantum field theory treats fields as defined over all of spacetime, not just at one moment, reinforcing the idea of a four dimensional geometry that includes time as a dimension of equal footing.

Experimental evidence in quantum theory, such as tests of Bell's inequalities by John Clauser, Alain Aspect, and others, shows that quantum entanglement and correlations do not respect classical notions of local realism. While not a direct proof of eternalism, these findings underscore that classical pictures of time and space must give way to subtler understandings. A block universe provides one possible metaphysical foundation that can accommodate such counterintuitive effects without requiring a global now.

**Practical Technologies and Their Implications**

One of the strongest indirect validations of relativistic spacetime structure, and thus the block universe notion, comes from practical technology. The Global Positioning System (GPS) is a prime example. GPS satellites orbit Earth and carry atomic clocks that must be synchronized with clocks on the ground. Special relativity predicts that the satellites' motion causes their clocks to tick slower relative to those on Earth. General relativity predicts that the weaker gravitational field at the satellites' altitude causes their clocks to tick faster. Careful calculations accounting for both effects are necessary to achieve the high precision that allows GPS to provide location information accurate to within meters.

Without incorporating relativistic corrections, GPS would quickly accumulate large errors. The fact that our global navigation systems require time dilation and the geometric view of spacetime to work correctly in practice is a strong endorsement of relativity's foundational claims. It indirectly supports the notion of a block universe by showing that treating time differently for different observers is not just theory but a practical necessity.

## Historical Progression and Conceptual Implications

The journey from Newton's absolute time to Einstein's relativity and beyond has been guided by experiments and observations. Newton's universe allowed a universal present and a consistent sense of time flowing. The 19th century attempts to detect the ether failed, leading to the constancy of the speed of light as a puzzle. Einstein's special relativity resolved this puzzle by merging space and time into a unified structure. Minkowski's geometric interpretation consolidated that view, while general relativity reinforced it by linking gravity to spacetime curvature.

By the mid 20th century, experiments repeatedly confirmed relativistic predictions, leaving little doubt that time and space are interwoven. Particle physics, cosmic ray observations, and high precision tests of gravitational effects have all validated the relativistic framework. Cosmological discoveries, including the CMB and detailed surveys of distant galaxies, have shown that what we label as past events remain imprinted in the light we see now, consistent with a four dimensional tapestry rather than a universe evolving along a universal timeline.

Quantum developments added complexity, but also allowed interpretations that fit into a block universe, whether through many worlds or relativistic quantum field theories. Over time, what began as a radical departure from common sense has matured into a robust theoretical and empirical framework. Eternalism and the block universe interpretation do not stand in isolation. They align with a century of accumulated evidence and theoretical refinements.

**For Example**

Consider a grand library representing all events in the universe. Each book on a shelf corresponds to a particular event at a particular time. In Newtonian physics, you might imagine that only one book at a time is real, and that as time passes, you move from one book to the next, with old ones disappearing and new ones being written. In a block universe, the entire library of books exists at once. You do not create or erase books, you simply change your reading position.

This metaphor is supported by the evidence we have reviewed. The library does not rely on a universal highlight that picks out the "current" book, because simultaneity and the definition of now differ from reader to reader (observer to observer). The library's structure is fixed, and events are all placed definitively on the shelves. Observers, following their own trajectories through spacetime, experience different sequences of books, giving rise to different notions of simultaneity and different measures of time intervals.

**The Intellectual Satisfaction of Alignment with Physics**

One of the key attractions of the block universe perspective is the intellectual satisfaction it provides. Physics strives for simplicity, elegance, and universality in its descriptions. By treating time as a dimension, we achieve a unified geometric framework that accounts for all known relativistic effects without resorting to artificial constructs or absolute frames of reference. Instead of fighting the counterintuitive aspects of relativity, eternalism embraces them, providing a metaphysical interpretation that matches what we measure and calculate.

Detractors might argue that the absence of a universal flow of time is too alien to reconcile with everyday experience. Yet everyday experience also resisted the idea of Earth orbiting the Sun or the notion that atoms are mostly empty space. Science's track record shows that reality can differ drastically from naive intuitions. The block universe concept is another instance where the scientific worldview encourages us to transcend common sense. The reward is a coherent vision that ties together cosmological observations, quantum field theories, gravitational phenomena, and practical engineering.

## Summary

The observational evidence and theoretical support for the block universe are not derived from a single source, but from a convergence of multiple lines of inquiry over more than a century of scientific progress. The invariance of the speed of light, confirmed by the Michelson Morley experiment and countless subsequent tests, triggered Einstein's formulation of special relativity. This led to the recognition that time and space form a four dimensional continuum without a universal now. Experimental demonstrations of time dilation and the relativity of simultaneity cemented the idea that observers traveling at different velocities slice the block universe differently.

General relativity's geometric interpretation of gravity as curvature in spacetime further supports a unified four dimensional model. Observations in astrophysics and cosmology, from Hubble's law of cosmic expansion to the CMB data, fit naturally into a block universe framework where past and future epochs coexist in the spacetime structure. Quantum theories, despite their probabilistic nuances, can also be reconciled with eternalism when viewed through appropriate interpretations or relativistic quantum field formulations.

Technologies like GPS, which require relativistic corrections to function accurately, offer pragmatic evidence that treating time as a dimension is not just philosophically pleasing but practically necessary. This underscores that the block universe perspective is not an academic diversion but a worldview intimately connected to how we understand and navigate the world.

While these observations and theories do not force us to adopt eternalism as the only philosophical interpretation, they make it a highly natural one. Each piece of evidence, from the earliest relativity experiments to the latest gravitational wave detections, chips away at the classical notion of a universal present and a flowing time. The consistency, simplicity, and explanatory power of the block universe picture stand on a century of accumulated scientific progress.

With the empirical and theoretical foundation now firmly established, we are better positioned to explore the more ambitious implications of eternalism. The chapters ahead will consider how the block universe interacts with emerging theories, advanced cosmological questions, philosophical puzzles about navigation through spacetime, and the nature of existence. Armed with this evidence, we move forward with a clearer understanding of why the block universe is not just a speculative idea, but one deeply rooted in the fabric of modern physics.

# Chapter 6

## **Mechanisms for Navigating the Block (Theoretical)**

In previous chapters, the block universe hypothesis was explored from historical, philosophical, and scientific angles. This perspective, in which all points in time coexist within a four dimensional spacetime continuum, emerges naturally from the theoretical framework of relativity. It aligns with certain metaphysical interpretations of time that reject the notion of a universal present. Accepting that the past, present, and future may all exist simultaneously raises a provocative question: if these events are laid out in a four dimensional structure, can one navigate through them in a manner analogous to moving through space?

The human experience of time is linear and unidirectional, defined by a clear arrow pointing from past to future. Although one can move freely in space, there is no accepted mechanism for traveling through time at will. Travel into the future at an accelerated rate (through relativistic time dilation) is possible in principle and is essentially routine for high speed particles, yet true bidirectional navigation, particularly into the past, remains a more challenging and speculative concept.

This chapter explores various theoretical mechanisms that physicists and mathematicians have proposed for navigating the block universe. These ideas include closed timelike curves derived from general relativity, wormhole solutions that might connect distant points in spacetime, gravitational manipulations of geometry, and more speculative quantum gravity approaches. While none of these mechanisms has been realized or confirmed experimentally, they shed light on the boundaries of physical law and the complexity of spacetime. Although these proposals appear extraordinarily difficult, if not impossible, to implement in practice, examining them clarifies how modern physics restricts or potentially admits exotic forms of temporal navigation.

## Closed Timelike Curves and the Geometry of Spacetime

One of the most direct routes toward navigating time is suggested by the geometry of spacetime itself in Einstein's General Theory of Relativity. The fundamental equations of general relativity, known as Einstein's field equations, allow for solutions known as closed timelike curves (CTCs). A closed timelike curve is a path through spacetime along which an object could travel forward in time (as it normally does) and eventually return to an earlier point in its own history. If such a curve existed and could be accessed, it would constitute a kind of loop in the temporal dimension, potentially enabling travel to the past.

Historically, interest in CTCs began with the work of Kurt Gödel in 1949. Gödel found an exact solution to Einstein's field equations describing a rotating universe that admitted CTCs. In Gödel's universe, an observer could, by traveling along a suitable trajectory, loop back to an event in their own past. Gödel's solution was a theoretical tour de force, showing that relativity did not explicitly forbid time travel. However, this solution required a universe with unusual global rotation properties, not one resembling our own expanding, nearly isotropic and homogeneous cosmos.

Other solutions have also been proposed. The Van Stockum dust solution, discovered earlier, and the Tipler cylinder, proposed by Frank Tipler in the 1970s, describe infinite rotating cylinders of matter that could theoretically create conditions where CTCs form. In these scenarios, if one could travel around the cylinder with sufficient speed, one might return to an earlier moment in time. These constructions remain hypothetical and rely on infinite or near infinite length cylinders of massive density, which are not physically observed. Moreover, they depend on unrealistic distributions of matter and energy.

The existence of CTCs raises severe conceptual problems. They allow paradoxes that challenge causality, such as the grandfather paradox, in which a time traveler could potentially prevent their own ancestor's existence. Physicists and philosophers have debated how to resolve these paradoxes. Stephen Hawking, for example, proposed the chronology protection conjecture, suggesting that the fundamental laws of physics, perhaps through quantum effects, would prevent the formation of CTCs in any physically realistic scenario. While the conjecture is not proven, it offers a plausible explanation as to why we do not observe such loops.

No experimental evidence supports the existence of CTCs in our universe. Observations of cosmic structure, the cosmic microwave background, and the distribution of galaxies do not indicate global rotations or matter distributions capable of producing closed timelike loops. For now, CTCs remain mathematical curiosities that test the logical and physical limits of general relativity, rather than offering a practical means of navigating the block.

## Wormholes as Potential Bridges

Wormholes, often associated with science fiction, are hypothetical tunnels connecting two distant regions of spacetime. They appear as solutions to Einstein's equations under certain conditions. First proposed in a modern context by Albert Einstein and Nathan Rosen in 1935 (leading to the notion of Einstein-Rosen bridges), wormholes initially arose as a theoretical link between two separate points in space. Later works by Michael Morris, Kip Thorne, and Ulvi Yurtsever in the 1980s explored the idea that if one mouth of a wormhole were moved at high speed or placed in a strong gravitational field, time dilation effects could cause one end to age differently than the other. If an observer entered the younger mouth, they might emerge at the older mouth, effectively traveling into the past.

The idea that wormholes could function as time machines relies on extreme conditions. To prevent the wormhole from collapsing, exotic matter with negative energy density is often required. Negative energy density is not a substance readily available in nature. Although the Casimir effect in quantum field theory shows that negative energy densities can occur between closely spaced conducting plates, scaling this up to macroscopic, stable amounts required for a wormhole is beyond current technology and may be fundamentally impossible.

Moreover, even if negative energy matter could be harnessed, engineering a stable wormhole would be a monumental task. Stability analyses suggest that small perturbations might cause catastrophic collapse. The energy requirements for manipulating such a structure would dwarf anything conceivable with known technology. Attempts to understand quantum field behavior in wormhole spacetimes also hint that quantum fluctuations might destroy their exotic properties, preventing their use as time machines.

To date, no observation hints at the existence of wormholes. Searches for gravitational lensing events that could indicate wormhole passages have come up empty. The detection of gravitational waves has revealed merging black holes and neutron stars, but not the signatures of stable wormhole throats. Wormholes remain theoretical constructs that stretch our imagination. They are valuable as thought experiments that probe the elasticity of spacetime and reveal how far physical law might be extended, but they do not currently offer a realistic mechanism for navigating the block.

**Manipulating Spacetime with Gravitational Fields**

Both CTCs and wormholes rely on complex and extreme geometries of spacetime. Another route to consider is the deliberate manipulation of gravitational fields. Since general relativity ties gravity to the curvature of spacetime, controlling mass-energy distributions could in principle shape the structure of time paths available to observers. Could advanced civilizations, far beyond our current capabilities, rearrange stellar masses or harness black holes to create regions where time loops become accessible?

To consider this, one must appreciate the enormous scale of gravitational engineering. Human technology allows the launching of satellites and the construction of particle accelerators, but shaping gravitational fields on a macroscopic scale would require moving masses comparable to stars or at least massive planets. Even positioning large masses precisely to form desired spacetime topologies is difficult to imagine. The energy scales involved exceed anything currently possible by factors of billions or trillions.

Some theoretical analyses have focused on rotating black holes, known as Kerr black holes. The Kerr solution to Einstein's equations describes a spinning black hole with a region near its horizon where frame dragging occurs, twisting spacetime. In principle, these regions might admit unusual causal structures. However, venturing into these environments is lethal. The tidal forces, extreme radiation, and instability of the geometry ensure that any attempt at using these regions for time navigation would be suicidal for any physical traveler. Moreover, detailed analyses often show that quantum effects, gravitational backreaction, and the inevitable formation of singularities or horizons prevent stable, accessible CTCs or time-travel conduits from forming.

Thus, while the mathematics does not forbid the idea outright, the engineering requirements and survival challenges are insurmountable. It is akin to suggesting that one might rearrange entire galaxies to produce desired time loops. Such theoretical musings highlight the immense gap between what the field equations permit in principle and what physical reality allows in practice.

**Quantum and Advanced Theoretical Approaches**

Modern physics has yet to produce a complete theory of quantum gravity. Such a theory is expected to unify general relativity, which governs large scale structure, with quantum mechanics, which dominates at the smallest scales. Candidates include string theory, loop quantum gravity, asymptotically safe gravity, and causal set theory, among others. Some of these approaches imply that spacetime may not be a smooth continuum at the Planck scale (approximately $10^{-35}$ meters) and may instead have a discrete or emergent structure.

In speculative discussions, if spacetime is emergent from underlying quantum degrees of freedom, perhaps manipulating those fundamental constituents could alter the arrow of time or create shortcuts. For example, quantum fluctuations might, under incredibly rare conditions, produce temporary structures resembling wormholes or enable information exchange across different epochs of the universe's history. If that were true, a sufficiently advanced civilization or a future experimental setup might exploit these quantum aspects to navigate the block.

These notions remain distant from current experimental reach. No known experiment can probe the Planck scale directly, and quantum gravity remains a theoretical frontier. Even identifying clear observational signatures of quantum gravity effects is challenging. Without empirical guidance, suggestions of time manipulation through quantum gravity remain hypothetical scenarios illustrating the full range of what theorists consider.

Proposals like the Alcubierre warp drive solution, while primarily considered for faster than light spatial travel, show that exotic spacetime metrics can arise in general relativity. Extending such ideas into the temporal domain would require even more significant amounts of exotic energy. If quantum gravity allowed stable negative energy states at macroscopic scales, this could open a door to engineering unusual causal structures. Yet, every indication suggests that negative energy conditions are tightly constrained and vanish quickly due to quantum fluctuations, effectively preventing large scale exploitation.

In addition, interpretations of quantum mechanics like the Wheeler-Feynman absorber theory once entertained time symmetric boundary conditions, where advanced and retarded waves interacted. This time symmetry hinted at a universe where future and past played symmetric roles. However, reconciling this with macroscopic causality and thermodynamics remains problematic. Even if microscopic processes are time symmetric, the macroscopic arrow of time restricts how these symmetries manifest at larger scales. Thus, no concrete approach emerges from these quantum considerations that would permit an observer to freely traverse the block.

**Information Through Time?**

A less ambitious notion than traveling bodily through time is the possibility of sending information backward or forward in a controlled manner. If signals could be transmitted to the past, it would constitute a form of temporal navigation. While the sender does not physically move, they could influence earlier events, effectively navigating the block by altering previously fixed slices of spacetime from the vantage of the future.

However, attempts to send information backward run into fundamental obstacles. The known laws of physics, from classical causality to quantum no communication constraints, prohibit the sending of usable information into the past. Quantum entanglement does not allow faster than light signaling or backward time communication because it only creates correlations, not direct causal influence. All known interactions preserve causality and prevent changing previously recorded outcomes.

If a hypothetical mechanism allowed backward communication, paradoxes arise again. The recipient in the past could act on the information to prevent it from being sent, undermining logical consistency. These causal paradoxes mirror those encountered with CTCs and strongly suggest that nature forbids such possibilities. Without a resolution to these paradoxes, even sending information backward remains a theoretical dead end. The block might be fully laid out, but the laws of physics seem structured to prevent rearranging its events once they are fixed.

## Engineering Challenges and Conceptual Limits

Even if tomorrow a new theory emerged showing that negative energy fields, stable wormholes, or quantum gravity manipulations were physically possible, the engineering challenges would be beyond formidable. Harnessing energies comparable to those found in supernovae or gamma ray bursts would be minimal compared to the task of shaping spacetime at will. Civilizations would need control over matter and energy on cosmic scales and technologies to stabilize delicate configurations against quantum instabilities.

Additionally, unknown principles could arise in a complete theory of quantum gravity that impose strict cosmic censorship on time travel attempts. Similar to chronology protection, these principles might ensure that any attempt to engineer a time loop triggers gravitational collapse, black hole formation, or quantum decoherence that prevents useful manipulation. The universe may have self defense mechanisms that preserve the integrity of causality and prevent temporal engineers from rewriting history.

Conceptually, even establishing a laboratory test for these phenomena poses deep challenges. How would one verify that a wormhole leads to a different point in time without sending a probe and confirming its return? Any experimental apparatus would have to withstand gravitational tides, extreme radiation fields, and intense quantum fluctuations. The practical impossibility of such experiments leaves these ideas as theoretical exercises rather than near future research projects.

**Theoretical Value - Despite Practical Impossibility**

The absence of practical methods for navigating the block universe does not render these theoretical explorations useless. On the contrary, exploring the boundaries of what relativity and quantum theory allow is a valuable intellectual endeavor. By probing these frontiers, physicists learn about the structure of laws that govern the universe, test the internal consistency of theories, and explore how different principles interact.

For instance, analyzing CTCs teaches physicists about the importance of causality and how delicate the causal structure of spacetime is. Studying wormholes reveals the complexity of maintaining stable geometries and the fundamental role of energy conditions in general relativity. Attempts to integrate quantum field theory in curved spacetime and to investigate negative energy densities improve our understanding of quantum vacuum states and the limits of energy conditions.

These considerations can guide the search for quantum gravity. If a future theory rules out all time travel mechanisms on a fundamental level, that would tell us something deep about the nature of spacetime. Conversely, if certain loopholes appear, they would demand new principles to handle causal paradoxes. Either outcome enhances our conceptual framework.

Additionally, these thought experiments have cultural and educational value. They help communicate the strangeness and subtlety of relativistic spacetime to broader audiences, stimulating curiosity and emphasizing that human intuitions about time are not always aligned with the underlying physical reality. Even if the block universe remains a static four dimensional structure that cannot be navigated freely, contemplating possible navigation expands our appreciation for the coherence and elegance of the laws that confine us.

## Beyond Physics: Philosophical Considerations

From a philosophical perspective, the difficulty or impossibility of navigating the block aligns with certain interpretations of eternalism. Eternalism states that all moments in time exist, yet it does not promise any accessibility beyond our ordinary forward progression. One might imagine the block universe as a vast landscape, but humans and other physical observers traverse a single path, unable to deviate, backtrack, or leap forward arbitrarily. This non navigability does not disprove eternalism, it merely underscores that existence does not guarantee reachability.

The combination of relativity and thermodynamics provides a natural explanation for why we are confined to a single temporal direction. The arrow of time, defined by the growth of entropy and the accumulation of memory records, ensures that even if the future exists in the block, we cannot bring knowledge of it into our current state without following normal causality. Eternalism thus remains consistent with the irreversibility of macroscopic processes, providing no simple handles for rewriting events.

Philosophers have also considered that the impossibility of time travel could maintain moral and existential coherence. If past events are fixed and cannot be altered, then accountability and responsibility remain meaningful. One cannot excuse one's actions by promising to correct them later through time manipulation. The block universe may contain all outcomes, but the laws prevent rearranging these outcomes at will, preserving a stable moral and historical fabric.

**Conclusion**

The notion that all events in time coexist in a block universe leads to an inevitable question: if they exist, can one navigate among them? Despite a century of progress in relativity, quantum theory, and cosmology, no known mechanism allows for free navigation through time. The solutions that hint at such possibilities, including closed timelike curves and wormholes, require conditions that appear to be physically unattainable. They invoke exotic matter, incredible gravitational engineering, or conditions on a cosmic scale that have never been observed.

General relativity's field equations may not strictly forbid these phenomena in principle, but the overwhelming weight of theoretical and observational constraints suggests that nature protects the timeline from interference. Hawking's chronology protection conjecture, while not proven, encapsulates the idea that time travel scenarios are either forbidden by unknown laws or are so unstable and dangerous that they cannot be realized.

Quantum gravity remains an uncharted territory. Future breakthroughs might shed more light on whether time navigation is absolutely excluded or if there exist subtle quantum effects that allow limited manipulation. Until then, the consensus is that traveling to the past, revisiting personal histories, or reshaping the future are beyond the reach of known physical principles.

The block universe may represent reality accurately, but reality itself, as governed by entropy increase, causal structure, and energy conditions, restricts observers to a linear trajectory from what we label the past to what we label the future. No advanced technology currently envisioned, no rearrangement of gravitational fields, and no quantum trick proposed so far has cracked the formidable fortress of causality and thermodynamics.

In the chapters that follow, the focus can turn to the broader philosophical and existential ramifications of eternalism without the promise of free navigation. Understanding that the block is fixed while we remain confined to a single worldline enriches the philosophical landscape. It presents a universe that is timeless in structure but tightly constrained in how we experience and influence the flow of events. If the block is a timeless tableau, it seems we remain part of it without the key to roam freely through its corridors.

76

# Chapter 7

## The Science of Closed Timelike Curves and Wormholes

Previous chapters have traced how the block universe hypothesis emerges from Einstein's relativity and how eternalism offers a philosophical foundation that treats time as a dimension on equal footing with space. These discussions raise intriguing questions: Can the geometry of spacetime itself admit structures that disrupt our intuitive sense of temporal order? Solutions to Einstein's field equations sometimes suggest the possibility of closed timelike curves (CTCs), which would allow an object to return to its own past, and wormholes, hypothetical tunnels that connect distant points in spacetime. Although these concepts often appear in speculative science fiction, they arise naturally, at least mathematically, from general relativity's equations.

This chapter examines the science behind CTCs and wormholes. It explores why these exotic solutions matter, how they challenge the standard chronology of events, and what theoretical and physical constraints might prevent them from existing in the real universe. Central to this inquiry are questions about causality, energy conditions, the role of exotic matter, and potential safeguards like Hawking's chronology protection conjecture. By understanding how physics approaches these remarkable geometric possibilities, one gains deeper insight into why nature might impose strict limitations that maintain a coherent and stable temporal order.

## Defining Closed Timelike Curves

In everyday experience, time appears to follow a single direction. Events occur in sequence, and one cannot revisit the past except through memory. In relativity, objects follow "world lines" through spacetime. These are paths traced out as time progresses from birth to death. Normally, world lines never loop back; they run from earlier to later times.

A closed timelike curve, however, represents a dramatic departure. A CTC is a world line that loops back to an earlier time. If an object traveling along such a curve could survive the journey, it might arrive at a temporal coordinate corresponding to its own past. This suggests a form of time travel that is not just metaphorical or partial, but literal: the traveler returns to an earlier point in its own history.

From a mathematical standpoint, CTCs emerge from certain solutions of the Einstein field equations, which relate mass-energy distributions to spacetime curvature. These solutions do not inherently forbid world lines from bending back on themselves in time. Yet, the existence of CTCs would have far-reaching physical and philosophical consequences, provoking questions about causality, consistency, and the fundamental nature of reality.

## Gödel's Rotating Universe

One of the earliest exact solutions demonstrating CTCs was published by Kurt Gödel in 1949. Gödel's universe is a highly idealized rotating cosmological model that differs dramatically from our observed universe. In Gödel's scenario, the presence of a global rotation sets the stage for paths in spacetime that loop back in time.

While this solution does not match empirical observations, no large-scale rotation or such exotic structure is found in cosmic surveys, Gödel's contribution was crucial. It proved that general relativity, by itself, does not forbid the existence of time loops. The Einstein equations, so successful in describing gravity and cosmic expansion, also allow for bizarre geometries that challenge our intuitive understanding of time's arrow.

Gödel's solution thus introduced a profound tension. Relativity's core does not inherently uphold a principle preventing world lines from returning to their past. Something else must forbid this, whether it is additional physics, quantum corrections, or energy condition constraints. Gödel's universe stands as a theoretical beacon illuminating a potential loophole in our understanding of temporal order.

## Tipler Cylinders and Other Exotic Constructions

Subsequent theoretical analyses produced other examples of CTCs. One famous construction is the Tipler cylinder, proposed by physicist Frank Tipler. A Tipler cylinder would be an infinitely long, massive, rotating cylinder of matter. Under carefully chosen conditions, infinitely long extension and extremely rapid rotation, it could theoretically induce regions of spacetime that form closed timelike loops.

Such solutions are often considered "bare existence proofs." They show that in principle, by carefully assembling mass and energy in certain configurations, one might obtain time loops. Yet these setups remain wildly unrealistic. Infinite cylinders of perfect matter or precisely tuned cosmic rotations do not appear in nature. Even slight deviations from the idealized conditions tend to break the formation of CTCs.

The main value of these examples is conceptual. They indicate that time loops are not forbidden by a naive reading of relativity. To rule them out or render them irrelevant, physicists must consider additional constraints like energy conditions, quantum effects, or global topological properties that differentiate physically plausible scenarios from mathematical curiosities.

## The Grandfather Paradox and Causality Issues

CTCs raise immediate and troubling issues for causality. Consider the grandfather paradox: if one could travel back in time and prevent one's grandfather from meeting one's grandmother, thus preventing one's own birth, a logical inconsistency arises. Who made the trip back if you were never born?

Classical physics depends on consistent causal ordering. Effects follow causes, and paradoxes do not arise. CTCs seem to undermine this principle, threatening the coherence of physics and logic itself. Various resolutions have been proposed. Some researchers suggest that self-consistency constraints must hold, ensuring that any attempt to alter the past fails in a manner that prevents paradoxes from materializing. Under such constraints, everything a time traveler does in the past was always part of history, leaving no room for genuine contradiction.

While such theoretical resolutions exist, they are uncomfortable. They impose restrictions that feel contrived, suggesting that an unseen hand enforces consistency. This discomfort motivated physicists to consider whether nature has more fundamental ways of preventing the formation of CTCs entirely.

## Chronology Protection Conjecture

Stephen Hawking introduced the chronology protection conjecture in the early 1990s. He argued that even if classical solutions to the Einstein field equations allow CTCs, realistic physics might forbid them. According to this conjecture, quantum field effects and other high-energy processes would arise near any forming CTC, generating infinite energy densities that destroy the would-be time loop before it stabilizes. In essence, the universe protects its own chronology.

This conjecture, while not proven, enjoys strong support in the physics community. It elegantly removes the paradoxical consequences of CTCs by showing that attempts to create them trigger processes that shut them down. Chronology protection thus points to a deep principle that may lie beyond classical general relativity, involving quantum effects or other fundamental constraints that prevent nature from realizing these bizarre configurations.

If chronology protection is correct, then CTCs, while mathematically admissible in simplified models, never manifest in real astrophysical or cosmological settings. This would resolve the tension raised by Gödel's universe and Tipler cylinders, restoring the causal order that physical consistency demands.

## Wormholes – Bridges Through Spacetime

Turning from CTCs to wormholes, one encounters another exotic consequence of relativity. Wormholes are hypothetical structures that connect distant regions of spacetime, functioning as tunnels or bridges. Initially introduced by Ludwig Flamm and further explored by Einstein and Rosen (leading to what was once called an Einstein-Rosen bridge), wormholes first appeared as purely theoretical constructs linking two black hole solutions.

For decades, wormholes remained a curiosity. In the 1980s and 1990s, work by Michael Morris, Kip Thorne, and others explored whether wormholes could be made traversable. If negative energy matter could stabilize a wormhole throat, perhaps travelers or signals could pass from one mouth to the other. This would offer a shortcut through space, potentially connecting two regions separated by immense distances.

Crucially, if one mouth of a wormhole could be moved at high speed or placed in a strong gravitational field and later reunited with the stationary mouth, time dilation could induce a temporal offset between the wormhole entrances. This configuration might transform a wormhole into a time machine: stepping through it could lead not just to another place in space but to another era in time.

**Wormholes as Time Machines**

The wormhole time machine concept captures public imagination because it aligns neatly with science fiction themes. However, the conditions required are daunting. Traversable wormholes that remain stable under perturbations demand exotic matter with negative energy densities that can counteract the natural tendency of gravity to pinch off any tunnel.

The exotic matter requirement is severe. Known forms of matter have positive energy densities. Negative energy appears transiently in quantum phenomena like the Casimir effect, but only in tiny, carefully arranged setups. Scaling this up to maintain a stable macroscopic wormhole throat is far beyond current understanding. Without a supply of stable negative energy, wormholes collapse before anyone could use them.

Moreover, even if negative energy were available, one must consider stability under quantum fluctuations, the buildup of radiation, and the gravitational backreaction caused by attempted travel. Analyses suggest that wormholes, if they can be created at all, would be incredibly sensitive to disturbances. Any attempt to use them as time machines might cause them to destabilize, either collapsing or degenerating into something unusable.

Like CTCs, wormhole-based time travel also raises causal paradoxes. The same logical inconsistencies arise if one can emerge from a wormhole mouth into an earlier time. Chronology protection may play a role here too, ensuring that quantum or gravitational effects prevent the realization of such scenarios.

**Energy Conditions and Exotic Matter**

Both CTCs and wormhole solutions run into constraints known as energy conditions. In classical general relativity, these conditions were introduced to ensure physically reasonable matter distributions. They assume energy densities are nonnegative and that matter behaves well under gravitational collapse.

To generate and maintain wormholes or unusual CTC geometries, one often needs violations of these conditions. Negative energy densities or exotic stress-energy tensors appear mandatory. Yet, no known stable, large-scale source of negative energy exists. While quantum field theory allows for short-lived negative energy fluctuations, these vanish quickly and cannot be harnessed at will.

This tension between what mathematics allows and what physical laws and energy conditions permit underscores why no observational evidence supports wormholes or CTCs. If something in the fundamental structure of matter and energy forbids stable negative energy on large scales, then wormhole time machines remain a nonstarter.

## Black Holes, Kerr Geometry, and Ring Singularities

Astrophysical black holes are the nearest real-world analog to exotic geometries, but they offer no convenient time loops or stable wormholes. Rotating black holes, described by the Kerr solution, have intricate internal structures. Deep inside the event horizon of a Kerr black hole, theoretical analyses suggest there might be ring singularities and regions where causal structure becomes complex.

However, no known process allows observers to survive journeying into these realms. Tidal forces and extreme conditions would tear apart any traveler. Furthermore, realistic astrophysical processes differ from the pristine mathematical Kerr solution. Collisions with infalling matter, nontrivial magnetic fields, and quantum effects likely eradicate any delicate causal anomalies. Thus, while black holes test the limits of spacetime geometry, they do not provide a practical route to producing or observing CTCs or wormholes usable for time travel.

## Quantum Gravity and Potential Resolutions

Many physicists believe that a complete quantum theory of gravity, yet to be fully formulated, will clarify whether CTCs and wormholes are genuinely possible or mere artifacts of incomplete theory. Quantum gravity approaches, like string theory and loop quantum gravity, attempt to unify relativity and quantum mechanics at the Planck scale.

If spacetime emerges from a more fundamental discrete structure, topological changes like creating wormholes or loops in time might be strictly forbidden. Alternatively, if they are possible in principle, they could be extremely rare or unstable, disappearing almost as soon as they form. Without experimental guidance, these remain hypotheses.

Quantum gravity could introduce selection rules or energy bounds that rule out negative energy configurations necessary for stable wormholes. Or it might reveal that any attempt to create a CTC triggers quantum gravitational fluctuations that restore causal order. In either case, the expectation is that a deeper theory refines or discards the problematic solutions admitted by classical relativity.

**Experimental Hints and Indirect Tests**

No direct observational evidence confirms the existence of CTCs or traversable wormholes. Searches in astronomy for exotic lensing signatures that could indicate wormhole mouths have not yielded results. Gravitational waves detected so far come from inspiraling black holes and neutron stars, with no hint of unusual topologies.

The absence of observational support does not prove impossibility, but it strongly suggests that if such structures exist, they are either incredibly rare, heavily shielded, or demand conditions not found in ordinary cosmic environments. Given the complexity of energy requirements and stability issues, most physicists suspect that nature is not hospitable to such shortcuts in time or space.

Indirectly, the very consistency of cosmological observations and the lack of obvious causal anomalies support chronology protection. The universe appears well-ordered over billions of years of evolution. If CTCs or wormholes that enable time travel were abundant, their effects might have manifested in unexpected correlations or anomalies in the data. Finding none strengthens the case for stable causal order.

**Philosophical Implications**

The theoretical existence of CTCs and wormholes touches on deeper philosophical issues. If time is a dimension in a block universe, these structures would, in principle, permit stepping outside the linear narrative we associate with temporal flow. The possibility of meeting one's earlier self or sending signals into the past undermines the intuitive arrow of time and the uniqueness of cause preceding effect.

Yet, the very fact that these scenarios seem so difficult to realize physically may illustrate how nature enforces consistency. Temporal order may emerge not just as a convenient approximation, but as a fundamental principle reinforced by the laws governing energy, matter, and quantum effects. Eternalism and the block universe may provide a conceptual framework where all moments exist, but physical accessibility to those moments could remain tightly constrained.

Philosophers and physicists alike find this comforting, as it preserves logical coherence, moral responsibility, and the intelligibility of historical narratives. If time loops and wormhole time machines never manifest, we need not worry about paradoxes or altered timelines. Instead, we gain confidence that causality, as we experience it, reflects a robust feature of physical law rather than a fragile artifact.

**The Role of Thought Experiments**

Considering CTCs and wormholes, even if they remain purely theoretical, has intellectual value. Such thought experiments push theories to their limits. By exploring the consequences of exotic solutions, physicists test the internal consistency of frameworks like general relativity and probe the interfaces with quantum mechanics.

These scenarios highlight areas where our current understanding may be incomplete. If classical relativity allows bizarre solutions, but nature forbids them, then something must refine the theory's domain of applicability. Chronology protection and quantum gravity research aim to identify these refinements, ensuring that the theory produces a universe that matches observations.

In this sense, CTCs and wormholes serve as conceptual tools. They uncover tensions, guide the search for deeper principles, and help frame questions about what is physically meaningful versus what is merely mathematically possible. Without these challenging constructs, physicists might overlook important subtlety in the structure of spacetime and the role of fundamental laws in shaping temporal order.

**Summary**

Closed timelike curves and wormholes represent some of the most extraordinary solutions admitted by Einstein's equations. By allowing loops in time or shortcuts through spacetime, they challenge our understanding of causality and the feasibility of time travel. Although classical general relativity does not forbid these structures outright, multiple lines of reasoning suggest that nature imposes stringent restrictions:

CTCs: Solutions like Gödel's universe and Tipler cylinders show time loops are possible mathematically. Yet, no known realistic configuration or astrophysical object produces them. The grandfather paradox and other causal dilemmas prompt theories like Hawking's chronology protection conjecture, implying quantum effects prevent CTC formation in practice.

Wormholes: Initially studied as bridges between black holes, wormholes gained attention when theorists realized that, with exotic matter, they might be made traversable. By manipulating relative motion, a wormhole could become a time machine. However, the need for negative energy matter and stability conditions, plus the causal paradoxes of time travel, strongly suggest that wormholes remain unrealizable fantasies. Observations reveal no sign of such structures, and their engineering demands lie far beyond current or anticipated capabilities.

Energy Conditions and Quantum Gravity: The extraordinary requirements for negative energy and the lack of any stable source for it support the conclusion that wormholes and CTCs are disfavored by fundamental physics. Hopes rest on quantum gravity to clarify these issues, and most physicists expect that the unified theory will further protect causality, excluding or rendering unachievable the wilder solutions of classical relativity.

Philosophical and Conceptual Value: While no practical paths to time travel emerge, studying these constructs sharpens understanding and highlights the robust nature of causal order. The absence of observational or experimental evidence strengthens confidence that time travel scenarios remain in the realm of speculation rather than reality.

In the broader narrative of understanding time, closed timelike curves and wormholes help delineate the boundaries of possibility. Their analysis reinforces the notion that, despite the block universe viewpoint, nature itself enforces temporal coherence. The result is a universe where all events may exist, but not all paths through those events are open to exploration, preserving the consistent temporal fabric upon which physics and human experience rest.

# Chapter 8

# Quantum Interpretations and the Many-Worlds Conundrum

The block universe hypothesis, with its vision of a four dimensional spacetime in which all events coexist, is rooted in classical relativity. Quantum mechanics, by contrast, introduces probabilities, superpositions, and measurement-induced phenomena that seem difficult to reconcile with a static, eternal structure. Classical thinking treats events as definite points in spacetime; quantum theory, on the other hand, deals in wavefunctions that represent multiple possibilities simultaneously, only yielding specific outcomes upon measurement.

Bridging these views is a nontrivial challenge. Interpretations of quantum mechanics vary widely, from the Copenhagen interpretation with its mysterious wavefunction collapse, to Bohmian mechanics with hidden variables, to relational and transactional interpretations, and finally to the many-worlds interpretation (MWI). Many-worlds, introduced by Hugh Everett in the 1950s, eliminates the notion of collapse altogether. Instead, it claims all possible outcomes occur in a vast branching structure of realities.

If the block universe suggests that all times exist, many-worlds suggests that all outcomes exist. Combining these ideas leads to the concept of a "block multiverse," a static yet unimaginably complex structure containing not only every point in spacetime but every possible quantum event and outcome as well. This chapter explores how quantum interpretations, especially many-worlds, intersect with eternalism and what that might mean for understanding time, causality, and the very nature of reality.

## The Measurement Problem and Wavefunction Collapse

Central to quantum mechanics is the measurement problem. Standard quantum theory uses a wavefunction to represent the state of a system, allowing for superpositions where particles can be in multiple states at once. However, when an observer measures the system, it appears to yield a single definite outcome. The Schrödinger equation, which governs wavefunction evolution, is deterministic and unitary, yet measurement outcomes appear probabilistic and non-unitary, suggesting some special process, often called "collapse", occurs during measurement.

The Copenhagen interpretation treats collapse as fundamental but unexplained. It says wavefunctions describe probabilities, and when observed, these probabilities crystallize into one actual result. Before measurement, an electron's spin might be both "up" and "down" in superposition. After measurement, one definite spin orientation emerges.

In a block universe setting, where all events are laid out, the notion of a sudden collapse is puzzling. If the future is already there, what does it mean for the wavefunction to collapse? If every event exists at once, including measurement results, where is the space for genuine uncertainty or probability?

## Many-Worlds Interpretation – No Collapse, Just Branching

The many-worlds interpretation aims to solve the measurement problem by abolishing collapse entirely. In many-worlds, the wavefunction describes not just probabilities but actual states that all occur. When a measurement takes place, the universe does not collapse to a single outcome. Instead, it branches into multiple non-communicating branches, each realizing one of the possible outcomes.

Hugh Everett's original proposal reframed the situation: the measuring device and observer also follow the unitary Schrödinger equation, becoming entangled with the system's outcomes. Instead of selecting one result, the universal wavefunction encompasses all results, each recorded by different versions of the observer in different branches. The randomness we perceive reflects our subjective ignorance of which branch we occupy post-measurement.

From a block universe perspective, many-worlds suggests that the four dimensional structure is not singular but part of a higher dimensional entity containing multiple timelines. At every quantum event, the block might branch into a suite of parallel worlds. There is no single future; instead, a vast array of futures spread out, all equally real. The classical block becomes a block multiverse, a static but high-dimensional structure containing every possible outcome of every quantum measurement.

### Eternalism Meets Many-Worlds – The Block Multiverse

Eternalism asserts that all points in time, past, present, and future, exist equally in a spacetime manifold. Many-worlds states that all outcomes of quantum events exist equally in a universal wavefunction. Combining these leads to a picture where not only does time form a static dimension, but the quantum state forms a branching tree of possibilities. Each branch is a distinct four dimensional block, complete with its own events. Multiply this branching across innumerable quantum events, and the structure resembles a colossal, static lattice of spacetime blocks.

In this block multiverse, time does not flow and outcomes do not get chosen. Instead, observers follow particular branches where events seem to unfold linearly and probabilistically. The branching is an illusion of perspective. From the global viewpoint, all branches and their outcomes sit side by side, timelessly coexisting. This synergy between eternalism and many-worlds can feel both elegant and unsettling. It removes collapse and absolute becoming, but introduces a reality so vast that it challenges our understanding of individuality, identity, and meaning.

**Quantum Superposition and Time – Conceptual Links**

Quantum superposition allows a particle to be in multiple states. Eternalism lets all times exist at once. Extending these ideas, one might speculate that the universal wavefunction encodes not only the positions and momenta of particles, but the entire geometry of spacetime and its possible configurations. In quantum cosmology and quantum gravity research, physicists attempt to write a wavefunction of the universe that includes gravitational degrees of freedom, potentially treating the shape and structure of spacetime as another quantum variable.

If successful, this approach might show that the block multiverse arises naturally from a more fundamental quantum description. The notion of branching worlds could be tied to how decoherence selects stable quasi-classical histories from a vast superposition of spacetime configurations. The familiar arrow of time and classical past-future distinction might emerge from decoherence, with observers finding themselves in stable branches that resemble a single timeline, even though globally, the structure is much richer.

## The Many-Worlds Conundrum – Probability and the Born Rule

While many-worlds avoids collapse, it introduces a challenge: why do we perceive probabilities as the Born rule suggests? The Born rule states that the probability of an outcome is given by the square of the amplitude of its corresponding wavefunction component. If all outcomes occur, what is probability measuring?

Many-worlds defenders propose that probability emerges from the measure of "branch thickness" or relative weight of branches. An observer splitting into multiple copies across branches will find themselves more likely in branches with greater amplitude weight. This is a subtle and debated issue, and no consensus exists on how to derive the Born rule purely from many-worlds principles.

In a block multiverse context, one might imagine that thickness measures how common certain outcomes are across the timeless landscape of branches. Your subjective experience of probability could arise from the fact that almost all observer-instantiations reside in branches with frequencies matching Born's rule. Although elegant in principle, the details remain a thorny conceptual problem, demonstrating that even a grand unification of eternalism and many-worlds does not solve every quantum puzzle.

## Decoherence and the Emergence of Classical Reality

A crucial ingredient for making many-worlds plausible is decoherence. Interactions with the environment cause quantum systems to lose coherence between different possible states. This splits the universal wavefunction into branches that no longer interfere. Each branch can then evolve as if it were a classical world with definite outcomes.

From a block perspective, decoherence can be viewed as the process that carves out distinct classical timelines from the quantum multiverse. Without decoherence, one would have an indescribably complex superposition of configurations. With decoherence, stable branches emerge, each corresponding to a well-defined history. Observers in these branches see a classical world, unaware of the parallel branches evolving alongside them.

In this sense, decoherence acts as a bridge between microscopic quantum uncertainty and the stable classical appearances we experience. It allows the block multiverse to be organized into effectively independent classical histories, each looking like a single block universe to its inhabitants, even though countless alternatives run in parallel at the global level.

## Implications for Navigating the Block – Infinite Complexity

In earlier chapters, navigating the block universe was already considered a formidable, likely impossible task. Introducing many-worlds only amplifies the complexity. Instead of one block, there is a vast multiplicity of blocks. To move backward in time, or to alter the past, would now mean contending with not just one timeline, but a branching structure of timelines. To navigate among these branches, one would need to somehow manipulate quantum states and undo decoherence, a feat beyond known physics.

No known mechanism allows an observer to jump between branches. Once decoherence has separated them, branches do not communicate. They are like different pages of a book that are never read simultaneously. The concept of navigating this block multiverse remains purely speculative and unsupported by any plausible physical model. If wormholes, CTCs, or other exotic constructs failed to provide navigation in a single timeline, they fare no better in a many-worlds scenario.

## Relational and Transactional Interpretations – Alternatives to Many-Worlds

Many-worlds is not the only interpretation that could inform how quantum mechanics fits into a block universe. Other interpretations also grapple with time and reality:

Relational Quantum Mechanics: Suggests that quantum states are always states relative to a particular observer or system. There is no absolute wavefunction collapse, only relational facts. In a block universe, events and outcomes might be seen as relative configurations among subsystems. This perspective eschews branching worlds, focusing instead on contextual relations that could integrate neatly with eternalism.

Transactional Interpretation: Introduces a time-symmetric view where advanced and retarded waves form a kind of "handshake" across time. In a block universe, since past and future coexist, a transactional approach that treats time symmetrically fits naturally. Yet it stops short of positing multiple realized outcomes, instead framing quantum events as standing waves that form when all conditions are met.

These alternatives provide different ways of thinking about quantum phenomena without invoking an infinite multiplicity of worlds. While they may not solve all interpretational problems, they offer conceptual tools that harmonize with the idea of a static, four dimensional spacetime and may help us think more symmetrically about time.

## Quantum Gravity and the Timeless Wavefunction of the Universe

Quantum gravity and quantum cosmology push these considerations further. In attempts to unify general relativity and quantum mechanics, some approaches eliminate time as a fundamental parameter, treating it as emergent from correlations in a timeless quantum state. The Wheeler-DeWitt equation, an attempt at a quantum gravitational Schrödinger equation, suggests a "wavefunction of the universe" that does not depend on time explicitly. Instead, time emerges from relationships between parts of the system.

If time is emergent and the block universe is an effective description at a certain scale, many-worlds might apply at a more fundamental level, describing a large quantum state with no preferred temporal slicing. Decoherence and low-entropy conditions in the early universe might then produce the arrow of time we experience, and yield stable branches that look like classical histories evolving in one direction.

In this picture, both eternalism and many-worlds might be emergent phenomena arising from a deeper quantum reality that is neither strictly time-bound nor classically deterministic. The block we observe could be one of many stable patterns of correlations in a timeless quantum substrate. This radical scenario remains speculative, but ongoing research in quantum gravity aims to clarify whether such visions have empirical consequences.

## Experimental Clues and the Nature of Reality

Interpreting quantum mechanics is notoriously difficult because all mainstream interpretations yield the same experimental predictions. Tests of Bell inequalities, weak measurements, and ever more precise quantum control experiments confirm quantum theory's predictions but do not favor one interpretation over another. Many-worlds, Copenhagen, Bohmian, and others all agree on measurable outcomes.

What experimentation can do is explore the boundary between quantum and classical realms. Quantum decoherence experiments, quantum computing, and advanced interference setups show how fragile quantum coherence is in macroscopic systems. These findings support the general story that reality emerges from quantum possibilities via decoherence. Yet, they cannot tell us if all outcomes occur, or if wavefunction collapse is "real."

Likewise, cosmological and astrophysical observations reveal a universe that looks classical at large scales, with well-defined histories recorded in cosmic microwave background patterns and galaxy distributions. If many-worlds and eternalism hold at the fundamental level, these classical features must emerge from deep quantum underpinnings that current observations do not directly probe. Perhaps future gravitational wave detectors or quantum gravitational experiments might hint at underlying quantum structures, but this remains speculative.

## Philosophical Reflections – Freedom, Identity, and Meaning

Combining many-worlds and eternalism can have profound philosophical implications. If every possible outcome of every choice exists in some branch of the block multiverse, what happens to responsibility, moral agency, or personal identity?

One might argue that if all outcomes occur, then individual decisions lose significance. Yet, from the perspective of any given observer in a single branch, outcomes still matter. You do not experience all outcomes simultaneously. Your branch is your reality, and your choices shape what you perceive next. The existence of other branches does not diminish the meaning of the path you follow. Your subjective experience remains consistent and fully real to you.

Personal identity becomes a more subtle question. Are the "you" in other branches the same person, or different individuals who share your past up to a branching point? Philosophers disagree. Some contend that identity is tied to continuity of experience, making each branch a new version of you. Others argue that if all future outcomes exist in parallel, the notion of a singular self is ill-defined. Perhaps identity is simply a stable pattern in the quantum multiverse, one that decoherence shapes but does not uniquely define.

Ultimately, these philosophical challenges highlight the difficulty of reconciling human intuitions with a reality that may be far stranger than classical thinking allows. The block multiverse scenario poses deep questions about free will, fate, and the role of observers in a grand cosmic tapestry where everything that can happen does happen, frozen in a vast configuration beyond time.

## Beyond Many-Worlds – Searching for a Unified Perspective

While many-worlds offers a neat solution to the measurement problem and a natural fit with eternalism, it is not the final word. Ongoing research in interpretations of quantum mechanics, quantum gravity, and the nature of time may produce novel insights. Hybrid interpretations, or entirely new frameworks, could emerge that treat time and quantum phenomena in unexpected ways.

For instance, quantum Bayesianism (QBism) interprets quantum states as personal degrees of belief for observers, avoiding the need for multiple worlds or classical collapse. In a block universe, QBism might say that the block includes all events, but the quantum state reflects an observer's internal knowledge rather than an objective multiplicity of outcomes. This perspective sidesteps branching structures and focuses on subjective probabilities.

Alternatively, future theories of gravity might show that what we call branching is just a mathematical convenience, or that time and quantum outcomes emerge from a deeper informational principle that does not require parallel universes. In this evolving landscape, many-worlds is a prominent contender, but it is not guaranteed to remain the most popular or coherent picture.

## Implications for Our Understanding of Time and Existence

If the block universe challenges us to see time as a dimension akin to space, many-worlds challenges us to see outcomes as equally real alternatives. Together, they paint a picture where what we call "now" and "actual" are just local illusions. The full reality might be timeless, containing every outcome, every history, and every future simultaneously.

This does not mean human life is meaningless or that choices lack importance. Rather, it reframes these concepts. Choice and meaning could be attributes of local branches within the grand structure, where observers care about their immediate experiences. The multiplicity of outcomes does not cancel the authenticity of one's particular storyline. Instead, it places individual narratives in a grander context.

By approaching quantum mechanics, time, and the universe in these broader terms, we gain intellectual flexibility. We learn to see that theories are models that must accommodate all observed facts, and that our intuitions, forged in a classical macroscopic world, may not be reliable guides to the fundamental structure of reality. The block multiverse scenario may be extreme and still hypothetical, but it encourages asking bold questions: Does reality have a preferred narrative, or are all narratives equally valid threads in a cosmic tapestry?

**Summary**

Quantum interpretations offer multiple avenues to reconcile the probabilistic quantum world with the static eternalism of the block universe. Among them, the many-worlds interpretation stands out for its synergy with eternalism. It eliminates wavefunction collapse and treats all outcomes as equally real, thereby extending the block universe into a branching block multiverse.

In this enriched view, time does not flow. Instead, decoherence carves out classical trajectories from a vast quantum state, and observers find themselves inhabiting particular branches. Probability becomes a statement about the measure of branches rather than a fundamental randomness. Although this scenario bypasses some conceptual difficulties, it introduces new ones, such as the interpretation of the Born rule and the meaning of individuality across infinite worlds.

Alternative interpretations, including relational and transactional perspectives, also aim to integrate quantum theory with a symmetric, timeless vision of reality. Quantum gravity research may further clarify these relationships, potentially revealing a deeply quantum and timeless underlying structure from which classical spacetime and probabilities emerge.

No matter which interpretation one favors, the interplay between quantum mechanics and eternalism expands our imagination. It suggests a reality that may be far more intricate and layered than a single, deterministic timeline. By acknowledging quantum branching or relational states, we come closer to understanding a universe where time, causality, and existence are woven together in patterns that transcend our everyday intuitions.

# Chapter 9

# Technological Challenges and the Quest for Exotic Matter

In previous chapters, we examined an array of theoretical frameworks that shape our understanding of the block universe. We discussed relativity, eternalism, quantum interpretations, and even many-worlds concepts. These ideas together paint a vision of spacetime as a vast, four, dimensional tapestry in which all events coexist. We also considered the bold notion of navigating this grand structure, not only traversing space but potentially venturing through time itself.

Moving from conceptual speculation to any practical attempt faces monumental hurdles. The laws of physics, as far as we currently understand, impose stringent conditions on manipulating spacetime. Chief among these challenges is the requirement for forms of matter and energy unknown to our present capabilities. Exotic matter with negative energy density emerges as a central theme. Such matter appears in theoretical discussions of traversable wormholes, time machines, and certain concepts akin to warp drives. All hinge on bending spacetime in ways impossible with ordinary matter.

This chapter examines the technological and physical barriers associated with engineering spacetime. We will clarify what exotic matter means, why it is deemed necessary, and how it relates to phenomena like the Casimir, effect. We will also consider the immense scale of energy and resources needed and assess whether our current scientific endeavors hint at any path forward. Ultimately, we will see how distant we remain from harnessing such matter, and what this implies for any realistic attempts at navigating the block universe.

## Why Ordinary Matter Will Not Suffice

When imagining altering spacetime, one might first think of immense machines, powerful beams of energy, or rearranging stars. While mass and energy can curve spacetime, as observed around stars and black holes, ordinary matter always contributes positive energy density. According to general relativity, spacetime curvature depends on mass, energy distributions. To create exotic structures like traversable wormholes or stable closed timelike curves, mere additions of normal mass and energy do not help. Adding more mass typically leads to gravitational collapse instead of forming usable bridges through spacetime.

Certain theoretical constructs demand negative energy densities to prevent spacetime from pinching shut. Negative energy can in principle create a repulsive gravitational effect, countering the natural inward pull of gravity and stabilizing a wormhole throat. Without such negative, energy configurations, wormholes would collapse instantly, much like trying to keep a tunnel open in shifting sand without support.

## Understanding Negative Energy Density

Energy density measures the amount of energy in a given region of space. Normally, this is positive. Negative energy density would mean a region of space has less energy than the surrounding vacuum. This energy deficit can have unusual gravitational effects, effectively creating a form of antigravity that opposes conventional curvature.

One well-known theoretical avenue to negative energy comes from the Casimir, effect. Proposed by Hendrik Casimir in 1948 and experimentally confirmed, the Casimir, effect arises when two conductive plates are placed extremely close in a vacuum.

Quantum fluctuations of electromagnetic fields are restricted by the plates, resulting in a measurable force pushing them together. In certain interpretations, the space between the plates possesses a lower energy density than the normal vacuum, effectively a negative, energy configuration.

This proof of principle shows negative energy is not pure fantasy. Nature allows it on tiny, controlled scales. Yet the negative energy from the Casimir, effect is minuscule and highly unstable. Scaling it up to macroscopic, or astrophysical magnitudes remains far beyond current technology. Still, it demonstrates that negative energy can exist, suggesting that with extraordinary advancement, larger manifestations might be conceivable.

**Exotic Matter - A Shopping List for Cosmic Engineering**

What properties would ideal exotic matter need to have to maintain a wormhole or enable a time machine?

**Substantial Negative Energy Density**: Enough to counterbalance intense gravitational collapse. Wormholes or spacetime tunnels demand tremendous negative energy to keep their geometry stable.

**Long-Term Stability**: Fleeting quantum fluctuations of negative energy are not enough. The exotic matter must persist and remain stable over potentially extended durations.

**Macroscopic Scale**: Achieving negative energy conditions at atomic scales is already difficult. For practical spacetime engineering, one must arrange these conditions at human, sized or spacecraft, sized scales.

**Control and Precision**: Even if negative energy were producible, it must be shaped and arranged with surgical precision. Random distributions of exotic matter would not form neat, traversable paths.

Meeting these criteria is extraordinarily challenging. We are essentially asking to sculpt vacuum fluctuations into stable, large structures that defy all conventional matter properties. This goes far beyond discovering a new alloy or superconductor. We are talking about engineering the fabric of spacetime itself.

**Energy Requirements - Beyond Any Known Capacity**

To curve spacetime enough for a wormhole or time travel scenario, one needs enormous energies. The gravitational fields produced by planets and stars cause relatively modest distortions. Creating a traversable wormhole or a stable CTC involves far more subtle and dramatic manipulation. The required negative, energy densities would demand energy resources beyond anything humanity can currently generate or even imagine.

Even if a method to generate negative energy were found, controlling it is another matter. Consider the delicacy of quantum states in experiments today. Achieving macroscopic quantum coherence is challenging enough; rearranging vacuum energy distributions on a large scale would be like sculpting incredibly intricate patterns in a turbulent quantum sea. Our existing energy production methods are insignificant compared to the scale needed. We might need to tap entirely new physical principles, perhaps zero, point energy or other exotic energy reservoirs, none of which are understood or harnessed at present.

## Hints and Early Steps - Quantum Optics and Metamaterials

Though we are far from engineering negative energy for spacetime manipulation, certain research fields provide glimpses of possible future directions:

**Quantum Optics**: Physicists manipulate quantum states of light and matter with extreme precision. Experiments have produced squeezed states of light, reducing quantum uncertainties in controlled ways. While not equivalent to negative energy densities, these show we can tailor quantum fields to some extent.

**Metamaterials**: Engineers have created materials with unusual electromagnetic properties, such as negative refractive indices. Although this relates to light propagation rather than gravity, it demonstrates that careful arrangement of structures at small scales can produce exotic responses.

**Ultra-Cold Atomic Gases and Bose - Einstein Condensates**: Cooling atoms to ultra, low temperatures yields macroscopic quantum states. These are not negative energy sources, but they teach us about controlling quantum states on larger scales than previously thought possible.

These lines of research are at best distant cousins of what would be required for negative energy engineering. Still, they hint that humans might continue to improve their capacity to manipulate quantum fields. Over centuries or millennia, incremental progress might reveal unexpected pathways. At present, these are just faint sparks, not even close to igniting the furnace needed for macroscopic negative energy production.

## The Stability Problem - Keeping the Tunnel Open

Assume, hypothetically, that we discovered how to generate negative energy on the required scale. Another significant issue arises: stability. Wormhole solutions often indicate extreme sensitivity to disturbances. Sending information or matter through a wormhole might cause quantum fields and gravitational configurations to fluctuate wildly, collapsing the structure.

Maintaining a stable wormhole would require a sophisticated feedback system constantly adjusting the distribution of exotic matter. Any spacecraft entering a wormhole would alter its internal fields, demanding real, time corrections. This would be analogous to trying to keep a delicate soap bubble intact while firing high, energy particles through it. Achieving such stabilization seems unimaginably complex.

## Paradoxes and Enforcement of Constraints

As discussed in earlier chapters, CTCs and time travel concepts raise severe causality paradoxes. If exotic matter and negative energy could enable stable time loops or wormholes, logical inconsistencies like the grandfather paradox could arise. This leads some physicists to believe that fundamental laws prevent the attainment of such configurations. Perhaps attempts to generate negative energy in bulk always fail or trigger instabilities. The universe might have built, in "chronology protection," ensuring that no advanced technology can break causality.

If so, the extraordinary engineering challenges we face are not just difficult, but inherently insurmountable. Negative energy might be technically possible in small doses, yet scaling it up to create workable wormholes or time machines could prove forever out of reach.

## Cosmic Laboratories and Astrophysical Phenomena

If we cannot produce exotic matter in a lab, could nature do it for us? Some speculative models suggest that in the early universe or deep inside certain hypothetical astrophysical objects, exotic energy conditions might fleetingly appear. If advanced civilizations exist, maybe they found ways to harness such rare environments. But so far, observational astronomy has revealed no stable wormholes or mysterious gravitational structures that would imply large, scale negative energy distributions.

Neutron stars, black hole accretion disks, and the interiors of compact objects push matter and energy to extremes, testing our theories. None of these environments show signs of negative energy phenomena that would hint at wormholes. Understanding these cosmic laboratories improves our knowledge of matter under extreme conditions, but it does not bring us closer to engineering exotic matter for spacetime manipulation.

## Philosophical and Practical Considerations

Given the daunting difficulties, it seems likely that block universe navigation will remain a philosophical or theoretical topic rather than a practical engineering project. We may debate time travel, wormholes, and eternalism much like we discuss parallel universes or the nature of consciousness. Without actual construction, these ideas remain in the domain of speculation.

Yet, considering these challenges refines our understanding of fundamental physics. Attempting to determine what is needed to manipulate spacetime at will pushes theories of relativity, quantum field theory, and energy conditions to their limits. Even if we never produce a gram of exotic matter, exploring these frontiers can yield new insights and perhaps drive the development of advanced mathematics, analog models, or conceptual frameworks that deepen our comprehension of the universe.

Moreover, acknowledging how difficult this is might reassure us that causality and temporal order are safe. If negative energy and exotic matter were easy to create, we might have observed unnatural gravitational phenomena or paradoxical cosmic signatures. Their absence bolsters the idea that the universe prevents such instability from arising naturally or through the efforts of advanced civilizations, if any exist.

## The Road Ahead - Incremental Progress or Acceptance of Limits?

Where do we go from here? Modern physics focuses on more modest and achievable goals: refining measurements of gravitational waves, exploring quantum computing to control quantum states for information processing, or searching for signs of new particles. None of these efforts aims at producing negative energy on a large scale.

Over extremely long timescales, if humanity endures and advances, we might uncover new states of matter or discover phenomena currently beyond imagination. Perhaps quantum gravity research will reveal ways to manipulate vacuum energy more directly. Maybe future civilizations, centuries or millennia from now, will look back at our skepticism as short, sighted. But with current knowledge, no clear path leads from present technology to harnessing negative energy for spacetime engineering.

There is also the possibility that fundamental laws impose absolute barriers. If every attempt at forming stable negative energy regions leads to catastrophic instabilities or logical contradictions, that might be the universe's final verdict. The block universe can exist as a theoretical construct, but physically rearranging its pages might remain forever forbidden.

## Comparison with Known Advanced Technologies

Consider known technological leaps: nuclear energy or space travel. Before the 20th century, these seemed far, fetched, yet they became realities once the underlying physics was understood and harnessed. The gap to exotic matter, however, is immensely greater. At least with nuclear energy, we had discovered atoms and understood their structure before building reactors.

In the case of negative energy, we do not even have a theoretical framework that suggests how to mass, produce and stabilize it. The Casimir, effect gives a microscopic hint, but no scaling strategy exists. The conceptual leap is more akin to asking if we can rewrite the rules of physics at will. Until we have a firmer grasp of quantum gravity and the vacuum structure, this remains pure speculation.

This stark difference emphasizes that some theoretical possibilities may remain out of reach, no matter how advanced we become. Nuclear power took decades from theory to practice, but at least it was built on well, tested atomic theory and known isotopes. Exotic matter stands on far shakier ground.

## Summary

The quest for exotic matter and negative energy is, in essence, the attempt to find a physical means to reshape spacetime dramatically enough to allow wormholes, CTCs, or other forms of block universe navigation. Current physics and technology find this goal utterly unattainable. Negative energy effects, though confirmed at quantum scales, remain minuscule and unstable. Scaling them up, stabilizing them, and molding them into a controllable geometry is beyond our present theories and engineering capabilities.

This challenge is not a matter of mere difficulty; it touches the bedrock of our understanding of the universe. Nature appears to guard the fabric of spacetime, preventing easy manipulation. The complexity, the need for precise negative, energy configurations, and the looming specter of causality paradoxes all suggest that even if the block universe is real, we cannot roam through it freely.

Nevertheless, exploring these ideas has value. By contemplating the requirements for exotic matter, we probe fundamental laws and refine our conceptual models. Such studies may lead to new insights or mathematical techniques that enrich physics and cosmology.

In conclusion, the technological challenges and the quest for exotic matter highlight the immense gulf between theoretical possibility and practical engineering. The block universe can be contemplated, understood intellectually, and appreciated philosophically, but forging a path through it remains beyond our current horizon. Exotic matter, negative energy densities, and stable spacetime sculptures might forever remain tantalizing impossibilities, testaments to the universe's intricate and unyielding design.

# Chapter 10

## Energy Requirements and Engineering the Impossible

Envision standing at the base of a towering cliff, its summit shrouded in clouds. Scaling it with today's tools seems unthinkable, yet you can measure its height and map its contours, knowing in principle that a peak exists. This analogy fits our predicament when confronting the energy requirements for manipulating or navigating the block universe. Theoretical frameworks suggest that, in a distant technological future, it might be possible to control spacetime at the most fundamental levels. However, the energy demands surpass anything we have ever contemplated.

Previous chapters highlighted the necessity of exotic matter and negative energy densities to create stable wormholes or closed timelike curves (CTCs). No known mechanism can achieve these conditions on a macroscopic scale. Here, we concentrate on the energy aspect: How vast must our power sources be? How would we harness and direct such energy? Even if advanced civilizations existed millions of years ahead of us, could they realistically meet these requirements?

### The Scale of Known Energy Sources

Throughout history, humanity progressed from burning wood to employing fossil fuels, nuclear fission, and now renewable sources like solar and wind. Yet, the total annual energy consumption of our global civilization is negligible compared to cosmic scales. The Sun emits about $3.8 \times 10^{26}$ watts of power. Even capturing a tiny fraction of this output continuously poses daunting engineering challenges.

To manipulate spacetime for wormholes or time travel, we must think many orders of magnitude beyond star-scale energies. Shaping spacetime at will rivals the gravitational effects of black holes, neutron stars, and galaxy collisions. Simply put, the energy needed for such feats dwarfs all familiar terrestrial or even stellar power sources. We must imagine tapping energies comparable to entire galaxies, and doing so in a controlled, precise manner rather than relying on raw cosmic violence.

**Dyson Spheres and Stellar Engineering**

The Dyson sphere concept provides a starting point for imagining large-scale energy collection. A Dyson sphere, a hypothetical megastructure encircling a star to capture much of its output, would yield energy millions of times greater than our current global consumption. Even so, channeling a star's energy might still fall short of the needs for producing negative energy densities large enough to stabilize wormholes.

Mere energy quantity is not enough. Negative energy, as discussed, involves delicate quantum manipulations. Even if we had a Dyson sphere's worth of power, converting it into the precise forms needed to warp spacetime remains uncertain. Building a Dyson sphere is itself a distant dream. Going far beyond that, to harness multiple stars or even a galaxy's worth of energy, staggers the imagination.

**Galactic-Scale Energy Harvesting**

If a star's output is insufficient, what about entire galaxies, each containing hundreds of billions of stars? A Type III civilization on the Kardashev scale might control the energy of a galaxy. Such power could, in principle, fuel attempts at spacetime manipulation. Yet, how to convert raw galactic energies into the intricate quantum states required for negative energy remains unknown.

Galactic power might be harnessed by extracting rotational energy from supermassive black holes, directing beams of relativistic particles, or orchestrating colossal gravitational lenses. Even then, this energy must be refined and focused into incredibly delicate configurations. It is not just about volume; it is about quality and precision. The complexity of such operations is beyond our current theoretical and engineering scope.

## Quantum Vacuum Engineering at Unimaginable Scales

The Casimir effect gives a tiny glimpse of how quantum fluctuations can produce negative energy densities. Two plates placed close together alter the vacuum state. Scaling this phenomenon to macroscopic, gravitationally significant levels would demand unimaginable resources.

Picture a structure spanning trillions of kilometers, lined with devices that tweak vacuum fluctuations at the Planck scale. Thousands of advanced machines, all orchestrated in perfect harmony, could attempt to "refine" the vacuum. Such a process would require energy and computational resources on a colossal scale, likely tapping stable artificial black holes or other exotic astrophysical objects as power sources. Every step here depends on unknown physics and engineering far beyond anything we can envision today.

## Stabilizing Structures Against Cosmic-Scale Forces

Assume we somehow generate negative energy fields. Another challenge looms: stability. Wormholes and other exotic configurations would be exquisitely sensitive to disturbances.

Any attempt to send matter or information through them could cause quantum and gravitational fluctuations that collapse the structure.

Maintaining stability would require dynamic feedback systems, immense computational networks that monitor conditions across astronomical distances, issuing corrections at unimaginable speeds. Constructing and powering such computational substrates introduces another layer of energy requirement. The energy needed to run these networks and prevent overheating, given their astronomical scale, compounds the primary challenge. Each engineering solution introduces further energy demands.

**Harnessing Exotic Astrophysical Objects**

Rather than building negative energy configurations from scratch, a future civilization might try to harness natural anomalies. Perhaps rare topological defects or primordial wormholes formed in the early universe could provide a starting point. Even then, energy would be required to stabilize and control such anomalies.

This scenario might reduce the brute-force energy cost of manufacturing exotic conditions but would still involve prodigious power expenditures. Securing, maintaining, and manipulating a naturally occurring wormhole would demand delicate quantum gravity engineering and constant energy infusion. The natural rarity of such phenomena suggests that few, if any, civilizations would have such luck, and even if they did, mastery of advanced physics and engineering would remain essential.

## Thermodynamic Constraints and Entropy Management

Energy usage on these scales must respect the second law of thermodynamics. Creating ordered states of negative energy or stable wormholes likely involves dumping waste heat and entropy elsewhere. How to radiate away the waste heat from a galaxy-scale vacuum engineering operation?

Vast cosmic radiators might be needed to prevent heat buildup. The complexity of managing heat flow at these scales adds another dimension to the energy problem. Entropy considerations mean that as you create conditions for wormholes, you simultaneously generate disorder elsewhere. Balancing this equation might require even more energy, more structures, and more ingenious solutions.

## Comparison with Theoretical Warp Drives

The Alcubierre warp drive proposal shows that even modest spacetime manipulation, faster-than-light travel without time travel, requires negative energy and enormous power. Some calculations indicate energy scales near that of entire planets or stars. If such a "simple" application of negative energy is already beyond our reach, the requirements for stable wormholes or controlled CTCs likely exceed that by orders of magnitude.

Studying warp drive proposals helps illustrate how sensitive these scenarios are. Any hope of navigating the block universe in a nontrivial way would likely be even more challenging than building a warp drive. This places block navigation solidly in the realm of near-impossible engineering feats.

## A Hierarchy of Engineering Challenges

We can imagine a hierarchy of technological feats:

**Current scale**: Building large particle colliders, fusion reactors.

**Planetary scale**: Harnessing all the energy of Earth.

**Stellar scale**: Constructing a Dyson sphere, capturing a star's entire output.

**Galactic scale**: Controlling energy from billions of stars.

**Spacetime engineering**: Manipulating negative energy, stabilizing wormholes, shaping quantum vacuum states.

The leap from stellar or galactic engineering to spacetime engineering is immense. It is not merely increasing quantity; it requires new physics and unprecedented complexity at every turn. The energy demands for block universe navigation likely reside at this highest rung, or even beyond it.

### Hope from Future Physics?

If new physics emerges, perhaps quantum gravity discoveries will reveal simpler ways to produce negative energy states. Maybe a trick exists that requires far less energy than brute-force approaches suggest. Such hopes hinge on major theoretical breakthroughs.

If vacuum energy could be tapped efficiently or if wormhole stability requires less negative energy than early estimates predict, advanced civilizations might find shortcuts. Still, pinning our hopes on future physics without any evidence is speculative. While possible, relying on unknown breakthroughs does not diminish the seriousness of the known energy barriers.

## The Ultimate Barrier or Just a Challenge?

From our current vantage, these energy requirements appear as ultimate barriers. However, we must remember that many past challenges once seemed insurmountable. Yet none of those past feats required rewriting the fundamental laws of physics or operating at cosmic scales. The energy needed for spacetime engineering transcends everyday engineering challenges.

If there are extraterrestrial intelligences millions of years older than us, could they have achieved this? Perhaps, but we see no empirical evidence. This silence might mean one of two things: either no civilization has reached these heights, or those that have operate beyond our detection. Another possibility is that even with infinite time and resources, the requirements remain too great.

One might argue that even if eternalism and the block universe are physically real, practical navigation might be no more achievable than altering the laws of physics themselves.

## Reconciling Theory and Reality

Eternalism and the block universe might be correct descriptions of reality's structure. The existence of wormhole solutions and CTCs in general relativity indicates that these are not forbidden by the simplest theoretical frameworks. Yet theory often neglects the full complexity of energy requirements, stability issues, thermodynamics, and quantum gravity effects.

In practice, the difficulty of producing negative energy densities and controlling them at large scales may effectively forbid block universe navigation. This is an example of how nature's practical constraints can be as decisive as its fundamental laws.

Just because something is mathematically possible does not mean it can be engineered. The universe's energy demands might serve as a de facto policing mechanism, ensuring causality and preventing timeline tampering.

**Conclusion**

The energy requirements for navigating the block universe stretch beyond any known or foreseeable engineering capability. Achieving the negative energy densities needed for stable wormholes or time travel would demand harnessing energies on galactic or even more extreme scales. We would need unprecedented precision in quantum vacuum engineering, massive computational infrastructure, and solutions to thermodynamic and stability issues. All these add up to a challenge so vast that it may never be met.

While this does not rule out eventual progress by future civilizations or the discovery of new physics that could ease the burden, nothing in our current understanding suggests an easy path forward. If anything, the enormity of the required energies reinforces the notion that the block universe, though theoretically conceivable, may remain practically inviolate. We can admire its grandeur, understand its theoretical underpinnings, and accept eternalism as a plausible interpretation of spacetime, but restructuring it to our liking appears, for now, and possibly forever, beyond our cosmic reach.

In the end, acknowledging these impossible energy demands emphasizes the distinction between what physics allows on paper and what nature permits in practice. The block universe stands as a timeless panorama, and any attempt at physically navigating it faces not just high hurdles, but a towering, cloud-shrouded cliff that may never be scaled.

## Chapter 11

## Spiritual, Ethical, Moral, and Existential Considerations

Having traversed the intricate theoretical landscapes of eternalism, quantum interpretations, and the staggering engineering challenges inherent in navigating the block universe, we now confront a pivotal juncture. The frameworks and monumental obstacles discussed thus far, while deeply insightful, may remain abstract and academic from our current standpoint. Nonetheless, the mere speculation about the possibility of altering time and space compels us to delve into profound ethical, moral, and existential questions that transcend pure physics and technology. This chapter explores the implications of such capabilities on human meaning, morality, social justice, identity, and the very essence of human existence.

### Rethinking Morality in a Non-Linear Temporal Landscape

In conventional moral philosophy, ethical responsibility is intricately tied to a linear conception of time. Actions committed in the present influence the future, and moral accountability follows the chronological progression of cause and effect. This forward-moving temporal structure underpins notions of justice, punishment, and reward. However, in a block universe framework where all points in time, past, present, and future, exist simultaneously, these foundational aspects of morality are profoundly challenged.

If beings possessed the hypothetical ability to traverse spacetime and alter past events, the ethical landscape would transform dramatically. For instance, suppose an individual uses a traversable wormhole to prevent a historical atrocity, such as genocide. On one hand, this intervention could save countless lives, embodying a utilitarian moral principle aimed at maximizing overall well-being. On the other hand, such an act could inadvertently erase events that led to subsequent societal advancements or moral awakenings spurred by adversity. The ethical calculus becomes tangled in a web of potential unintended consequences, making it difficult to evaluate the righteousness of interventions.

Moreover, the assumption of a non-linear temporal framework necessitates a reevaluation of moral theories. Deontological ethics, which emphasize duties and rules irrespective of consequences, might conflict with the fluidity of time manipulation. Conversely, consequentialist frameworks would need to account for the complex interplay of altered timelines and the cascading effects of temporal interventions.

Philosophers like Immanuel Kant and John Stuart Mill have long debated the foundations of moral responsibility, and the block universe adds a new layer to these discussions. Kantian ethics, with their focus on categorical imperatives, may struggle to accommodate actions that have retroactive implications. Similarly, Mill's utilitarianism would require a more sophisticated understanding of how outcomes are evaluated across multiple timelines.

The integration of these ethical theories with the block universe model reveals deep tensions. Traditional moral reasoning assumes a clear cause-and-effect chain, a linear progression where actions lead to consequences. In a block universe, where all events coexist, the boundaries between causes and effects blur, necessitating a more holistic and perhaps relativistic approach to ethics.

## The Grandfather Paradox as an Ethical Puzzle

Previously, the grandfather paradox was discussed in the context of causality and logical inconsistency within CTCs. However, this paradox also serves as an ethical puzzle. If an individual travels back in time and prevents their own grandparents from meeting, thereby erasing their own existence, the moral implications are profound.

From a deontological perspective, such an action would be inherently wrong, violating the duty to preserve human life and existence. However, from a consequentialist viewpoint, preventing one's own erasure might seem self-preserving, yet it raises questions about the broader impact on the timeline. The act of altering the past introduces a recursive ethical dilemma: the very means by which one ensures personal survival could result in the elimination of that survival.

This paradox extends to less personal scenarios. Consider a time traveler aiming to avert the rise of a tyrannical dictator. While the immediate ethical motive, to prevent suffering and loss of life, is clear, the long-term consequences could be morally ambiguous. Preventing a dictator's rise might save lives, but it could also remove the impetus for political and social reforms that arise in response to such authoritarian regimes. The ethical evaluation of time-altering actions requires a nuanced understanding of both immediate and long-term consequences, potentially conflicting with established moral principles.

Furthermore, the grandfather paradox underscores the potential for moral and logical inconsistencies inherent in time manipulation. It raises questions about the feasibility of ethical frameworks that rely on a fixed temporal sequence. If actions can retroactively negate their own existence or alter historical events in unpredictable ways, maintaining a coherent ethical system becomes exceedingly challenging.

Philosophers like David Lewis have explored such paradoxes, questioning whether consistent time travel is logically possible or whether it necessitates the existence of parallel universes to resolve inconsistencies. The ethical dimensions of these theoretical explorations compel us to reconsider the foundations of moral responsibility and accountability in a universe where time is not strictly linear.

**Power and Inequality - Who Controls the Timeline?**

In a scenario where time manipulation becomes feasible, the distribution of power would be radically redefined. Control over mechanisms that allow for the alteration of historical events or the preemption of future tragedies would centralize immense power in the hands of a select few, whether they be governments, corporations, or clandestine organizations.

Such centralization raises significant ethical concerns regarding inequality and abuse of power. A governing body capable of rewriting history could manipulate outcomes to maintain political dominance, suppress dissent, or engineer societies to fit specific ideological frameworks. The ethical dilemma centers on the potential for temporal tyranny: a state where power is not only exerted in the present but extends backward and forward across all of time.

Historical analogies, such as totalitarian regimes that seek to control narratives and rewrite history through propaganda, underscore the dangers of concentrated power. However, the temporal dimension amplifies these dangers exponentially. The ability to edit timelines could lead to a form of omnipotence, where temporal justice becomes a tool for perpetuating existing power structures rather than correcting injustices.

Moreover, the ethical question arises as to who gets to decide which timelines are altered. The potential for moral absolutism or relativism in these decisions could result in conflicting ethical standards, leading to international tensions or temporal conflicts. Establishing transparent, accountable, and equitable governance structures would be essential to mitigate the risks of such profound power.

Philosophers like Michel Foucault have examined how power dynamics shape societal structures, and the introduction of time manipulation technologies would add a new layer to these dynamics. The ethical imperative to prevent the misuse of temporal power aligns with concerns about surveillance, control, and autonomy in modern societies.

Additionally, the potential for a technological elite to monopolize time manipulation capabilities raises questions about social justice and equity. If only a privileged few can access and control such power, the resulting societal stratification could be unprecedented, with ethical ramifications that echo through multiple timelines.

### Identity, Free Will, and Meaning in a Timeless Universe

The block universe challenges fundamental aspects of personal identity and the concept of free will. In a traditional linear timeline, individuals make choices that influence their future, fostering a sense of autonomy and responsibility. However, if all points in time are fixed within a block, the notion of free will becomes ambiguous. Are our choices truly free if their outcomes are already embedded in the spacetime structure?

From an existential standpoint, this raises questions about meaning and purpose. If the future is already determined, do actions hold intrinsic value, or are they mere unfolding of a pre-existing script? This perspective can lead to existential nihilism, where the significance of individual agency is undermined. Alternatively, it could foster a sense of interconnectedness and acceptance, seeing oneself as part of a grand, unalterable tapestry of events.

Philosophers such as Daniel Dennett and Derek Parfit have explored these themes, debating the compatibility of free will with determinism. The block universe introduces a spatial dimension to this debate, complicating the relationship between choice and outcome. While some argue that free will can coexist with eternalism by focusing on subjective experiences of agency, others contend that it represents an illusion in a fixed spacetime structure.

Additionally, the concept of identity becomes more fluid. If multiple versions of oneself exist across different branches or timelines, what constitutes the "self"? Are these duplicates truly the same person, or distinct entities sharing a common origin? This multiplicity challenges traditional notions of singular identity and introduces ethical considerations about the treatment and rights of these multiple selves.

Philosophers like John Locke and Thomas Reid have long debated the nature of personal identity, and the block universe adds a new dimension to these discussions. The existence of multiple, coexisting selves could lead to a reevaluation of ethical obligations and moral duties, extending them beyond singular identities to a network of interconnected existences.

Furthermore, the block universe's timeless structure may influence existentialist thought, as discussed by philosophers like Jean-Paul Sartre and Albert Camus. The search for meaning, purpose, and authenticity becomes intertwined with the recognition of an eternal, unchanging reality, challenging individuals to find significance within a fixed temporal framework.

**Cultural Perspectives and Moral Relativism**

Different cultures possess unique moral frameworks, religious beliefs, and temporal understandings. Introducing the possibility of navigating the block universe and altering timelines would interact with these diverse cultural lenses in complex ways.

In some Eastern philosophies, such as Hinduism and Buddhism, time is often viewed cyclically rather than linearly. The idea of eternal recurrence or rebirth resonates with the block universe's timeless structure, potentially reinforcing or reshaping these beliefs. Time manipulation technologies might be interpreted as fulfilling spiritual prophecies or aligning with cyclical views of existence.

Conversely, Western philosophies and religions typically emphasize linear progressions of time, with clear beginnings and ends. The ability to alter the past or foresee the future could lead to theological debates about fate, divine providence, and human agency. Religious doctrines might either adapt to incorporate temporal manipulation or resist it as an overstepping of divine boundaries.

Moral relativism further complicates the scenario. If multiple timelines or branches coexist, each reflecting different cultural norms and values, establishing universal ethical standards becomes challenging. Conflicting moral principles across branches could lead to ethical pluralism, where no single moral framework holds supremacy, or moral absolutism, where some branches are deemed ethically superior.

The ethical implications of cultural diversity in a block universe extend to questions about imperialism, cultural preservation, and the right to alter or erase cultural histories. The potential for time manipulation to homogenize or fragment cultural identities necessitates a thoughtful approach to global ethics and intercultural dialogue.

Philosophers like Kwame Anthony Appiah and Alasdair MacIntyre have examined moral relativism and the role of culture in shaping ethical norms. In a block universe context, these discussions gain new dimensions as temporal manipulation intersects with cultural ethics, potentially requiring a reevaluation of moral universals and the acknowledgment of diverse temporal ethics.

**Legal Systems in a Time-Fluid World**

Law relies on historical precedent and the assumption of a stable, linear causality. Legal systems assign guilt and innocence based on actions that have occurred, relying on the continuity of events to establish accountability. In a block universe where timelines can potentially be altered, these legal foundations are upended.

If past events can be modified, the basis for legal responsibility becomes unstable. Crimes could be retroactively erased, or individuals could preemptively alter their past to avoid punishment. Legal systems would need to adapt to address these temporal ambiguities, potentially incorporating new laws that govern the ethical use of time manipulation technologies.

Legal theorists might propose temporal regulations, establishing jurisdictions over timeline alterations and setting boundaries for acceptable interventions. However, enforcing such laws presents unprecedented challenges. Traditional legal mechanisms, based on linear time, would struggle to accommodate events that can change retroactively.

Moreover, the question of evidence becomes more complex. If timelines can be edited, historical records might no longer be reliable, undermining the foundations of criminal justice. Establishing temporal integrity would require advanced monitoring systems, possibly involving temporal witnesses or time-stamping mechanisms that prevent unauthorized alterations.

Ethical and legal debates would likely arise about the rights to alter one's timeline, privacy across temporal dimensions, and the protection of historical integrity. The intersection of law and time manipulation introduces a new realm of jurisprudence, where temporal ethics must be woven into the fabric of legal theory and practice.

Philosophers like Ronald Dworkin and H.L.A. Hart have explored the nature of law and its relation to morality. In a block universe scenario, these theories would need to incorporate temporal dimensions, redefining concepts like justice, punishment, and rehabilitation to accommodate the fluidity of time.

## Collective Responsibility and Stewardship of History

The potential to alter timelines imposes a collective moral responsibility to steward history responsibly. Rather than viewing history as a static record, it becomes a dynamic resource that can be shaped for future benefit or detriment. This stewardship entails ethical decisions about which events to preserve, alter, or erase, with profound implications for collective memory and cultural heritage.

One approach to stewardship is minimal intervention, preserving historical integrity to maintain authenticity and continuity. This perspective emphasizes the importance of learning from the past, valuing the consequences of historical events as necessary for moral and societal growth. By maintaining an unaltered timeline, societies uphold the lessons and cultural developments that emerge from enduring past hardships.

Alternatively, a proactive stewardship model advocates for selective interventions to prevent suffering and promote well-being. This approach aligns with humanitarian ethical principles, aiming to eradicate atrocities and improve the overall quality of life across timelines. However, this model risks unintended consequences, as altering historical events could disrupt the delicate balance of societal evolution and moral development.

Philosophers like Hans Jonas have emphasized the ethical responsibility to future generations, advocating for a precautionary approach to technological advancements. In a block universe context, this translates into a responsibility to protect the integrity of historical timelines while navigating the potential for ethical interventions.

Effective stewardship would require robust ethical frameworks, transparency, and collective decision-making processes. It would necessitate a consensus on the principles guiding temporal interventions, ensuring that actions are aligned with broader societal values and do not favor specific groups or agendas.

Furthermore, the concept of temporal justice emerges as a critical consideration. Ensuring that all timelines receive equitable treatment, preventing the marginalization or exploitation of certain branches, would be essential for maintaining ethical integrity across the block universe.

**Compassion, Empathy, and Universal Moral Principles**

Core virtues such as compassion and empathy form the bedrock of ethical conduct. In the context of a block universe with the potential for timeline manipulation, these principles become even more critical. The ability to alleviate suffering or prevent harm through temporal interventions could be seen as the ultimate expression of compassionate ethics.

However, implementing such interventions responsibly requires a deep understanding of the broader consequences and an unwavering commitment to universal moral principles. Compassion must be balanced with foresight and wisdom to ensure that interventions do not lead to greater harm or ethical dilemmas.

Philosophers like Martha Nussbaum have emphasized the importance of empathy in ethical decision-making. In a block universe scenario, empathy would extend not only across present interactions but across temporal boundaries, fostering a sense of connectedness and responsibility towards all points in time.

Universal moral principles, such as the Golden Rule or the Ten Commandments, might serve as guides for responsible timeline manipulation. These principles advocate for actions that respect the dignity, rights, and well-being of all beings, regardless of temporal context.

Yet, the complexity of temporal ethics presents challenges in defining and applying universal principles consistently. The potential for conflicting moral imperatives across different timelines requires a nuanced approach to ethical reasoning, one that prioritizes overarching values while accommodating contextual variations.

Moreover, the application of universal moral principles in a block universe context necessitates a reconsideration of their scope and flexibility. How do these principles account for actions that transcend linear causality? What mechanisms ensure that compassionate interventions remain aligned with broader ethical standards across multiple timelines?

These questions underscore the need for a dynamic and adaptable ethical framework capable of addressing the multifaceted challenges posed by temporal manipulation. It calls for interdisciplinary collaboration between ethicists, philosophers, and scientists to develop comprehensive guidelines that uphold universal moral values while navigating the complexities of a block universe.

**Moral Education and Preparedness**

If the technological capacity for time manipulation were ever to be realized, it would necessitate a radical shift in moral education and societal preparedness. Educating future generations about the ethical implications of temporal interventions would be paramount, ensuring that individuals understand the profound responsibilities and potential consequences of such capabilities.

Moral education would need to integrate traditional ethical theories with novel considerations arising from the block universe and time manipulation. Curriculum development could include studies on temporal ethics, the philosophical implications of eternalism, and the practical challenges of maintaining moral integrity in a non-linear temporal framework.

Furthermore, fostering a culture of ethical mindfulness and responsibility would be essential. Societies would need to cultivate virtues like humility, foresight, and restraint, preparing individuals to handle the moral dilemmas posed by temporal manipulation technologies.

Scenario planning and ethical simulations could become integral parts of educational programs, allowing individuals to explore hypothetical interventions and their consequences in a controlled environment. These exercises would help develop critical thinking and moral reasoning skills necessary for responsible decision-making in complex temporal contexts.

Moreover, interdisciplinary collaboration between ethicists, physicists, engineers, and policymakers would be crucial in establishing comprehensive guidelines and regulations governing the use of time manipulation technologies. This collaborative approach ensures that ethical considerations are embedded into the development and deployment of such transformative capabilities.

The integration of ethical education with technological advancements serves as a safeguard against the misuse of time manipulation, promoting a balanced approach that respects both moral integrity and scientific progress. It emphasizes the importance of preparing society to navigate the ethical landscapes of a block universe responsibly, ensuring that temporal interventions, if ever possible, are conducted with profound ethical awareness and accountability.

## Embracing the Mystery and Limits of Control

Despite the allure of manipulating time and reshaping history, it is vital to recognize and embrace the inherent mysteries and limits of human control over spacetime. The block universe may offer a static, all-encompassing view of reality, but the ability to navigate and alter it introduces profound ethical and existential uncertainties.

Acknowledging the limits of control fosters a sense of humility and respect for the natural order of the universe. It reminds us that, while theoretical possibilities exist, the practicalities and ethical implications may render such capabilities beyond our moral and technological reach.

Moreover, embracing the mystery of the block universe encourages a focus on the present moment and the ethical use of the powers we currently possess. It emphasizes the importance of living ethically within the temporal constraints we navigate daily, rather than aspiring to godlike control over time and history.

This acceptance does not negate the pursuit of knowledge or technological advancement but grounds it within a framework of ethical responsibility and existential mindfulness. It calls for a balanced approach, where the quest for understanding does not overshadow the imperative to uphold moral integrity and respect the fabric of spacetime.

Furthermore, the philosophical notion of embracing uncertainty aligns with existentialist themes explored by thinkers like Søren Kierkegaard and Jean-Paul Sartre. It underscores the importance of finding meaning and purpose within the constraints of our temporal existence, rather than seeking to transcend them through speculative technologies.

## Conclusion

The possibility of navigating and manipulating the block universe brings to the forefront a myriad of spiritual, ethical, moral, and existential considerations. These issues extend far beyond the realms of physics and engineering, probing the very essence of human identity, responsibility, and the meaning of existence.

As we contemplate the theoretical frameworks and monumental challenges discussed in previous chapters, we must also grapple with the profound ethical implications of wielding such power. The ability to alter timelines, reshape history, and preempt future events poses ethical dilemmas that challenge our traditional moral paradigms. It raises questions about accountability, power distribution, identity, cultural preservation, and the fundamental nature of free will.

In navigating these complex ethical landscapes, it is crucial to develop robust moral frameworks, foster interdisciplinary collaboration, and prioritize virtues such as compassion, empathy, and responsibility. While the block universe may remain an abstract theoretical construct, its exploration serves as a valuable exercise in ethical reasoning and philosophical inquiry.

Ultimately, the block universe and the hypothetical technologies that could navigate it compel us to reflect on what it means to exist, to choose, and to act ethically in a reality where all moments coexist eternally. These reflections enrich our understanding of morality and identity, reminding us that even in the face of unimaginable possibilities, the principles that guide our actions and the meanings we derive from our lives remain paramount.

By confronting these spiritual, ethical, moral, and existential questions, we deepen our appreciation for the complexity and beauty of the universe, and affirm our commitment to living meaningfully and ethically within the temporal framework we inhabit. The journey through theoretical physics thus becomes intertwined with a profound exploration of the human condition, highlighting the inseparable relationship between our scientific aspirations and our moral imperatives.

## Chapter 12

## **Covert Applications - Government and Military Interests**

The prospect of navigating, influencing, or manipulating the block universe transcends mere scientific and philosophical debates, plunging into the realm of strategic and geopolitical considerations. History underscores that whenever groundbreaking technologies emerge, government agencies and military organizations are among the first to recognize and exploit their strategic potential. From the advent of nuclear energy to the development of artificial intelligence, states have consistently sought advantages that could secure superiority, deter adversaries, and control critical information. The hypothetical ability to navigate the block universe or rearrange events represents the pinnacle of strategic resources: it offers the unprecedented power to rewrite outcomes before they occur, anticipate adversarial moves with absolute precision, and engineer conditions that ensure perpetual dominance on a global scale.

In this chapter, we delve into the clandestine and covert dimensions of block universe navigation. We examine how secretive government agencies or paramilitary organizations might approach these possibilities, how they could internally justify their actions, the forms that covert operations might take, and the potential ways they might communicate, or fail to communicate, with the public. Additionally, we explore how such secret projects could inadvertently or deliberately manifest as unexplained phenomena in our skies and historical records, potentially offering explanations for some reported unidentified aerial phenomena (UAPs) that defy conventional physics.

## The Strategic Rationale

Understanding why governments and militaries would be interested in block universe manipulation requires an exploration of what such capabilities imply. Control over time is not merely another weapon in the arsenal; it represents a meta-weapon that transcends the conventional constraints of geography, resources, and linear cause and effect that govern traditional warfare. If an agency could access past events, it could preempt enemy research, sabotage foundational technologies before they are fully developed, or prevent influential leaders from rising to power. Conversely, it could ensure its own nation's prosperity by reinforcing certain historical choices, guaranteeing stable alliances, or steering societies toward beneficial technological and social paths.

For security strategists, even partial navigation of the block universe, such as limited backward observation for intelligence-gathering without active intervention, would revolutionize espionage. Imagine intelligence operatives who could witness historical diplomatic meetings first-hand, verify enemy weapons tests that were never officially recorded, or glean vital details from future scenarios to better prepare current defenses. This level of intelligence dominance would fundamentally alter the rules of global strategy, making it possible to anticipate and counteract adversaries' moves with unparalleled accuracy.

## Compartmentalization and Black Projects

In a world dominated by classified research, sensitive technologies are often developed under strict secrecy protocols known as "black projects." The manipulation of the block universe, if even remotely feasible, would constitute the blackest of all black projects. Governments would go to extreme lengths to conceal any hint of success or even the pursuit of such abilities. The reasons are clear: revealing even a fragment of this capability would trigger international alarm, spark intense competition, and destabilize the existing geopolitical order.

To maintain secrecy, agencies would likely compartmentalize research into isolated units, each working on seemingly unrelated problems. For instance, one unit might focus on quantum field manipulations, another on gravitational anomaly detection, and yet another on advanced vacuum energy experiments. Only a handful of top-level strategists and scientists would possess comprehensive knowledge of the project's entirety, ensuring that no single leak could expose the full scope of their endeavors.

This level of secrecy would extend to plausible deniability. Officially, governments would deny any interest in time travel or block universe navigation, dismissing such rumors as science fiction or conspiracy theories. Meanwhile, covert teams operating under innocuous departmental names, such as "Advanced Phenomena Assessment Unit", would push the boundaries of physics in hidden laboratories, possibly located in remote deserts, deep underground bunkers, or secured research vessels at sea.

**Exotic Matter and the Ultra-Secret Arms Race**

If exotic matter or negative energy densities are indeed required to manipulate spacetime significantly, securing such materials or production methods becomes a strategic imperative. States would likely invest billions of dollars into obscure advanced physics facilities, quantum computing centers, and zero-point energy research groups. Rival powers would engage in a race to unlock the secrets of stable negative energy pockets, akin to a new nuclear arms race but infinitely more subtle and profound.

Sabotage and espionage between nations would intensify as each power vies to crack the secrets of negative energy generation. Elite spy cells might infiltrate rival labs to steal experimental data or discredit scientists through smear campaigns. Double agents could introduce false theories to competitors, leading them down technological dead ends and wasting valuable resources. The stakes of such an arms race would be catastrophic: the first nation to achieve reliable time manipulation would enjoy a near-invincible position, capable of ensuring perpetual leadership by undermining its enemies' historical roots or erasing their future successes.

### Influence on World Events and Historical Outcomes

Consider a hypothetical scenario where covert agencies progress beyond mere observation and engage in limited temporal interventions. Even minor alterations in the past could yield significant strategic advantages. For instance:

**Case Study**: Operation Horizon

**Objective**: Prevent the rival nation's development of a groundbreaking missile guidance system in the 1980s.

**Method**: Utilizing a wormhole-like structure, agents introduce a subtle quantum perturbation to disrupt a chain of experimental successes in the rival's research laboratories. This perturbation causes a slight delay in the maturation of the technology, ensuring that the rival's missile program lags by three decades.

**Outcome**: In the present, the rival nation continues to struggle with missile technology, unaware that their historical hurdles were not natural but engineered by Operation Horizon. The initiating power gains decades of uninterrupted energy dominance and strategic advantage.

To maintain secrecy, all documentation and records pertaining to Operation Horizon would be compartmentalized. Agents involved would undergo rigorous security protocols, potentially including memory alteration techniques if such advanced methods are available, to ensure plausible deniability. Officially, documents would refer to "resource allocation corrections" and "historical anomaly management," never mentioning time manipulation or temporal interventions. Any historian or analyst discovering inconsistencies in the rival's research timeline would likely attribute them to mundane factors, such as administrative errors or natural delays, never suspecting covert temporal sabotage.

Such interventions, however, carry the risk of paradoxes and unintended consequences. Agencies might develop temporal simulation models to predict ripple effects, yet uncertainty would remain a persistent issue. This uncertainty would likely fuel internal debates within covert units, forcing agencies to adopt strict doctrines aimed at minimizing timeline disruptions. Interventions would be limited to carefully controlled scenarios, with comprehensive simulations and robust failsafes to prevent catastrophic outcomes.

## Ethical Conflict Inside Military Ranks

Not everyone within these covert structures would be comfortable with the concept of timeline tampering. Ethical dissidents, scientists, engineers, even military officers, might question the morality of rewriting events and erasing individuals' free will. Internal power struggles could emerge, with some factions advocating for full-scale temporal control while others push for a non-interventionist stance.

Whistleblowers, driven by moral convictions, might risk their lives to reveal partial truths about these secret projects. Such acts could lead to mysterious disappearances, discrediting campaigns, or professional ostracization. This internal tension could explain bizarre leaks or strange rumors that occasionally surface in the public domain. Occasional, cryptic insider revelations might allude to "projects involving rewriting the past," only to be dismissed by official spokespeople as conspiracy theories or science fiction.

These ethical conflicts within covert units underscore the profound moral dilemmas posed by temporal manipulation. Balancing strategic advantage with ethical responsibility becomes a delicate act, fraught with potential for internal dissent and moral compromise.

**Misattribution and Unidentified Aerial Phenomena**

One intriguing possibility is that covert agencies experimenting with spacetime manipulation might inadvertently produce anomalies observable to civilians. Imagine test flights of advanced craft designed to exploit subtle gravitational distortions or quantum thrusters powered by zero-point energy. These craft, unbound by conventional aerodynamics or inertia, could perform maneuvers that appear impossible: rapid accelerations, sharp right-angle turns, or sudden disappearances and reappearances.

Such sightings, reported by pilots and citizens alike, would be baffling and unexplainable by current aviation standards. They might be labeled as Unidentified Aerial Phenomena (UAPs) or UFOs. In reality, these could be experimental vehicles from clandestine programs attempting to understand or harness block universe properties. These advanced craft could be designed to probe weak points in spacetime, scout potential wormhole connections, or test negative energy fields in controlled environments.

Over decades, accumulated UAP reports often describe objects defying known physics, with erratic flight patterns and unexplained propulsion systems. If these reports correlate with secret test ranges or align with known black project locations, it could suggest a link between UAPs and covert temporal experiments. While extraterrestrial visitation remains a popular explanation for UAPs, another plausible theory is that these sightings are manifestations of humanity's own advanced technology derived from block universe research.

Moreover, some UAP sightings could represent time-travel reconnaissance drones, tiny, unmanned probes slipping between temporal layers. If an agency attempts to gather data from the future or the past, these probes might momentarily appear in present skies, causing confusion before blinking out. The public would witness strange lights or craft without ever realizing they were glimpsing experimental temporal probes.

**Covert Agreements and International Treaties**

If multiple global powers approach block universe navigation capabilities, secret pacts or treaties might form to manage the potential risks and prevent a temporal arms race. Just as international treaties regulate nuclear arsenals, covert agreements could set boundaries for timeline interference, ensuring that no single nation gains disproportionate temporal power.

These agreements, negotiated behind closed doors, might commit signatories to refrain from major historical edits, limit temporal surveillance to observational purposes only, or establish mutual oversight mechanisms, perhaps managed by an artificial intelligence designed to monitor and log any temporal alterations. The challenge lies in enforcing such treaties, given the secretive nature of these projects. Verification methods could involve quantum "reference beacons" placed outside normal spacetime, designed to detect timeline modifications and ensure compliance.

However, such cooperation is inherently fragile. How do you trust rival nations not to cheat when they can manipulate the very fabric of history? Mutual deterrence might stabilize a temporal arms race, creating a situation where no nation dares to alter timelines openly because of the fear of retaliation and the potential for catastrophic paradoxes. This uneasy equilibrium could shape global politics, with top leaders privately aware that the stability of reality rests on the restraint of all parties involved in temporal manipulation.

### The Intelligence Community's Role

Intelligence agencies, adept at deception and covert operations, would play a central role in controlling information about block universe projects. Counterintelligence teams would track suspected leaks, plant disinformation to mislead curious scientists, and produce fake research papers to camouflage genuine breakthroughs. These agencies might exploit cultural narratives, leveraging conspiracy theories, UFO folklore, or sensational rumors to cloak their real activities.

For example, if a scientist involved in a temporal project publishes a controversial theory, intelligence operatives could disseminate false critiques or alternative explanations to discredit the research. Similarly, if an experimental craft from a secret program is spotted, agencies could attribute it to natural atmospheric phenomena or optical illusions, diverting public attention away from the true nature of the technology.

Additionally, these agencies might utilize advanced psychological operations to shape public perception, ensuring that any anomalies or unexplained phenomena are misinterpreted as extraterrestrial activity or other benign explanations. This strategy maintains plausible deniability while allowing the covert programs to continue their temporal experiments without public scrutiny.

## Psychological Warfare and Societal Manipulation

Temporal capabilities would extend beyond traditional military power, opening new avenues for psychological warfare. Agencies could introduce subtle changes in the cultural narrative by altering historical events, removing certain influential philosophers, artists, or activists from history, thereby shaping ideological evolution. Instead of engaging in conventional warfare, conflicts might be fought by editing educational milestones, redirecting social movements before they emerge, or subtly influencing public opinion through temporal interventions.

Such manipulation poses significant ethical crises. The public, if ever uncovering the truth, might feel profound betrayal, questioning the authenticity of their cultural heritage and collective memory. Governments might justify these acts as necessary for maintaining stability or preventing catastrophic wars, even though the true motivations and outcomes of these interventions remain hidden.

## Inadvertent Disclosures and the UAP Connection

Sometimes, testing advanced gravitational or quantum-drive vehicles might go awry, resulting in unexpected public sightings. Locals living near test zones could report odd crafts defying known aeronautics. Official explanations might attribute these sightings to natural atmospheric phenomena, experimental aircraft prototypes, or optical illusions, ensuring that no direct connection is made to temporal manipulation projects.

As years pass, these sightings accumulate, and investigative journalists, ufologists, and curious citizens propose alien visitation theories to explain the unexplainable. However, agencies quietly ensure that the alien narrative protects their secrets. If a reporter uncovers advanced theoretical papers hinting at negative energy propulsion, a well-orchestrated disinformation campaign could discredit the findings, labeling them as fringe science or misinterpretations.

In this light, many UAP incidents might be misinterpreted evidence of humanity's secret endeavors to control time and spacetime. Lights in the night sky could be experimental platforms for scanning temporal structures, while unexplained radar returns might be artifacts of temporal distortions caused by attempts to stabilize wormholes or create temporal conduits. The public might see strange lights or craft but never connect them to the underlying temporal experiments conducted by covert agencies.

## The Moral Burden on Operators

Individuals executing these covert missions would bear an enormous moral burden. Scientists working in secret labs, aware that their breakthroughs could reshape history, might experience profound ethical dilemmas. Engineers and military officers directing time manipulation operations might question their loyalty and the moral justification of their actions, knowing that their efforts could erase entire branches of potential futures or alter historical outcomes in unpredictable ways.

To cope with these moral conflicts, agencies might implement psychological conditioning programs, rationalizing actions as necessary for national survival or the greater good. Morale officers and indoctrination programs could reinforce the notion that temporal interventions are paramount for preventing existential threats or ensuring societal stability. Over time, such justifications could create a closed moral universe within these secret communities, where the act of altering timelines is viewed as not just acceptable but essential for the nation's or civilization's prosperity.

Yet, cracks might appear. A rogue scientist or a conscientious military officer could leak subtle clues to the outside world, encrypted references in obscure scientific journals hinting at the existence of negative energy anomalies or temporal manipulation projects. These leaks might contribute to the swirl of UAP and conspiracy lore, leaving future generations with puzzles to decode and interpret, never fully grasping the true nature of these secret projects.

## Paranoia and the Constant Fear of Retaliation

If time manipulation capabilities become known or even suspected by rival powers, trust between nations would evaporate. Every historical event could be questioned: Was this war's outcome natural or engineered? Did that cultural renaissance occur spontaneously or was it nurtured by temporal interventions? Paranoia would spread within elite political circles, leading leaders to maintain temporal watchdog units tasked with detecting anomalies in historical records or monitoring quantum signatures indicative of timeline tampering.

This atmosphere of suspicion could paralyze decision-making processes, as leaders might hesitate to take bold actions for fear of unintended temporal consequences or retaliatory interventions by rivals. The knowledge that any nation could potentially rewrite history would create a fragile balance, where global stability depends on the unspoken restraint of all parties involved in temporal manipulation.

## Temporal Disarmament Negotiations

In a distant future, as the understanding of temporal manipulation deepens, calls for temporal disarmament might emerge. Just as international treaties regulate nuclear arsenals, temporal disarmament agreements would seek to set boundaries for timeline interference. These accords, negotiated behind closed doors, could commit signatories to refrain from major historical edits, limit temporal surveillance to observational purposes, or establish mutual oversight mechanisms, perhaps managed by an impartial artificial intelligence designed to monitor and log any temporal alterations.

Enforcing such treaties would be challenging, given the secretive nature of temporal projects. Verification methods might involve exchanging quantum keys that reveal subtle manipulations or employing quantum reference beacons placed outside normal spacetime to detect any unauthorized timeline changes. However, the effectiveness of these measures would be limited by the inherent secrecy and the ability of states to develop methods of plausible deniability.

Mutual deterrence might stabilize the temporal arms race, creating a situation where no nation dares to alter timelines openly due to the fear of retaliation and the potential for catastrophic paradoxes. This uneasy equilibrium could shape global politics, with top leaders privately aware that the stability of reality rests on the restraint of all parties involved in temporal manipulation.

**The Legacy of Secret Temporal Projects**

If one day the existence of secret temporal projects were to surface, through whistleblowers, declassified documents, or accidental disclosures, the revelation would fundamentally alter humanity's understanding of history and reality. A collective identity crisis would ensue, as individuals grapple with the knowledge that their history has been manipulated or altered by covert operations.

Rewriting textbooks to acknowledge temporal interventions would be traumatic, as societies confront the possibility that certain historical events were not organic but engineered. Philosophers, theologians, and ethicists would hold emergency summits to salvage meaning from a universe where history is malleable and contingent upon the actions of secretive agencies.

Unidentified Aerial Phenomena (UAP) research would undergo a seismic shift. Ufology conferences, once ridiculed, might gain newfound credibility as they explore the connections between UAP sightings and secret temporal experiments. Scholars would meticulously analyze old UAP reports, correlating them with known secret installations and declassified documents. The narrative of alien visitors could fade, replaced by a sobering acknowledgment of humanity's own hidden capabilities and moral dilemmas.

## Sociopolitical Fallout and Public Response

The public, upon learning that their governments engaged in temporal manipulation, would react with a mix of outrage, despair, and resignation. Protests could erupt, demanding the dismantling of all temporal technologies and the establishment of strict oversight mechanisms. Activist groups might form around demands for a "prime timeline charter," ensuring that no further manipulations occur and that historical integrity is preserved.

Leaders might face significant backlash, as citizens grapple with the betrayal of trust and the ethical implications of having their history altered without their knowledge or consent. The revelation could lead to a widespread loss of faith in government institutions, prompting demands for greater transparency and accountability in all areas of research and development.

Religious leaders might interpret temporal manipulation as an overstepping of divine boundaries, framing it as humanity "playing god" and violating sacred natural orders. Conversely, some might embrace it as fulfilling spiritual prophecies or aligning with cyclic views of existence, especially in cultures where time is seen as cyclical rather than linear.

Philosophers and ethicists would debate the morality of past interventions, discussing whether such actions can ever be justified by their outcomes. The ethical discourse would need to address the balance between preventing suffering and preserving the authenticity of historical development, ensuring that future moral progress is not compromised by temporal tampering.

## Cultural Reflections in Art and Media

Art, literature, and film would explode with representations of time manipulation's clandestine era. Fictional accounts might mirror the complexity of these covert projects, crafting narratives where hidden agents traverse centuries, subtly shaping civilization's trajectory. These stories would explore themes of power, responsibility, and the ethical dilemmas inherent in controlling time.

In visual arts, artists might create works that depict fractured timelines, paradoxical scenes, and the ethereal beauty of wormhole structures. Literary works could delve into the psychological and moral conflicts faced by individuals aware of their ability to alter history. Films might portray secret organizations battling to control temporal technology, highlighting the moral costs of wielding such power.

Immersive virtual reality experiences could simulate the sensations of navigating the block universe, allowing users to witness historical events from multiple perspectives or explore the consequences of hypothetical temporal interventions. These cultural reflections would serve as both a means to process collective trauma and a cautionary tale about the ethical responsibilities of wielding immense power.

## Post-Disclosure Governance and Ethics

In the wake of disclosure, where the existence of temporal manipulation projects is acknowledged, new global institutions might form to oversee and regulate the use of such technologies. A "Block Ethics Council," composed of philosophers, scientists, religious leaders, and citizen representatives, could debate and establish guidelines for responsible temporal interventions.

These institutions would aim to ensure that any use of temporal manipulation aligns with universal moral principles, prioritizing the well-being of all individuals and societies over narrow national interests. They might develop comprehensive ethical frameworks that address the complexities of altering timelines, emphasizing transparency, accountability, and the preservation of historical integrity.

The transition from secrecy to public stewardship would mark a significant shift in global governance. It would require unprecedented levels of international cooperation and ethical consensus, ensuring that temporal technologies are used responsibly and ethically. This shift would also involve rigorous oversight mechanisms to prevent misuse, with stringent penalties for unauthorized temporal interventions.

## The Ongoing Mystery of UAPs

Even after all revelations, some UAP sightings might remain unexplained. While many could be attributed to secret temporal projects and advanced covert technologies, others might represent genuine unexplained phenomena. The duality of attributing some UAPs to human endeavors and others to natural or extraterrestrial origins keeps the mystery alive.

This ongoing ambiguity ensures that the study of UAPs continues to intrigue scientists, enthusiasts, and skeptics alike. It also underscores the limitations of our current understanding, highlighting the need for continued research and open-minded exploration of both terrestrial and cosmic anomalies.

## Conclusion: A World Forever Changed

The exploration of covert applications of block universe navigation reveals a staggering horizon of complexity, where scientific possibilities intersect with profound ethical and geopolitical implications. Secret government agencies and military organizations, driven by strategic imperatives, could harness temporal manipulation as the ultimate strategic resource, reshaping history and securing dominance in ways previously confined to the realm of science fiction.

The secrecy surrounding these endeavors would necessitate extreme measures to maintain plausible deniability, including disinformation campaigns, compartmentalized research units, and the exploitation of cultural narratives to obscure the truth. The ethical dilemmas inherent in temporal manipulation would challenge traditional moral frameworks, raising questions about responsibility, power distribution, and the very nature of free will and identity.

If the existence of secret temporal projects were ever revealed, the sociopolitical fallout would be profound, shaking the foundations of trust between governments and citizens. Public discourse would be inundated with debates about the morality and legitimacy of altering timelines, while cultural expressions in art and media would grapple with the complexities of temporal ethics and the human condition.

Ultimately, the chapter underscores that the block universe concept is not merely an abstract theoretical construct but a potential pivot around which power, ethics, and identity revolve. Governments and militaries, by venturing into these shadows, would redefine what it means to govern, secure, and guide a civilization. The consequences of their actions, once hidden, would challenge every assumption we hold about who we are and the nature of the reality we inhabit.

In contemplating these covert applications, we recognize that the interplay between advanced technology and ethical responsibility is as intricate and multifaceted as the fabric of spacetime itself. The hypothetical scenarios presented serve as a cautionary tale, urging us to consider the profound moral implications of wielding power that can reshape the very essence of history and human existence.

## Chapter 13

## Unidentified Aerial Phenomena and Temporal Anomalies

For decades, people have reported strange objects in the sky, sometimes performing impossible maneuvers or appearing and disappearing without explanation. Such sightings, variously called UFOs, UAPs, or anomalous aerial vehicles, have long puzzled the public, scientists, and government officials. Initially dismissed as misidentifications, hoaxes, or illusions, UAPs have gained renewed attention as pilots, military personnel, and credible witnesses describe crafts that violate known aerodynamics and physics.

In earlier chapters, we explored the theoretical possibility of manipulating block universe coordinates and the covert agencies that might attempt such feats. Here, we connect those threads directly to UAPs. Could some UAP sightings represent the byproducts of experiments with spacetime, the presence of temporal anomalies, or even reconnaissance probes from other block coordinates? As we delve into this topic, we'll consider how advanced block navigation technology might explain UAP characteristics, why sightings often correlate with sensitive locations, and how the public discourse around these phenomena may mask more profound underlying physics.

### The Challenge of Explaining UAPs with Conventional Physics

Modern aviation and aerospace engineering rely on well-established physical laws. Aircraft must generate lift, require propulsion systems, and obey inertia. UAPs, particularly those described as performing instantaneous acceleration, right-angle turns at high speeds, or hovering motionlessly without visible means of thrust, defy these principles.

For decades, skeptics and believers have debated explanations, ranging from secret military aircraft to extraterrestrial visitors. But each hypothesis struggles with key questions: How do these objects resist immense inertial forces during abrupt maneuvers? How do they remain completely silent at supersonic speeds? And why do they appear transiently and vanish, leaving no trace?

Temporal anomalies offer a new lens. If craft are not merely moving through space but also manipulating the local fabric of spacetime, many perplexing behaviors could be reinterpreted. By bending spacetime or exploiting pockets of negative energy, a vehicle might sidestep inertial stresses, appear to defy gravity, and "blink" in and out of visible ranges if it shifts slightly along the temporal axis.

**Time Distortions as a Propulsion Mechanism**

If a hypothetical craft can adjust local spacetime geometry, it might create effects analogous to a bubble of altered time flow around it. Inside this bubble, the craft experiences normal physics. To an external observer, however, the craft might seem to accelerate instantaneously because, from our vantage point, its internal timeline runs differently.

Imagine an observer watching a video in slow motion while the subject inside the video experiences normal speed. To the observer, the subject's movements might appear abrupt or discontinuous. Similarly, if a UAP harnesses temporal gradients, it could appear to "jump" from one position to another, as if teleporting or performing impossible maneuvers, when in reality, it's moving in a region of spacetime with altered flow rates.

Another possibility is that a craft using such technology can momentarily detach its trajectory from our timeline's standard progression. Instead of pushing against air and gravity, it travels through a slightly shifted temporal slice, encountering fewer obstacles or reduced inertial mass. Upon returning to our time frame, it appears to have leapt forward or changed direction effortlessly.

**Localized Temporal Anomalies and Optical Effects**

Strange lights, shimmering auras, or sudden appearances and disappearances could arise from local temporal distortions affecting how light reaches the observer. If a vehicle warps or bends spacetime, light rays passing near it might bend or refract oddly. Such distortions could produce optical illusions, making the craft seem elongated, contracted, or pulsating with color shifts as our perception grapples with a region where time and space metrics differ from normal conditions.

In some accounts, witnesses describe objects vanishing "as if someone turned off a switch." This abrupt disappearance could be explained if the UAP shifts its temporal alignment. By slipping a tiny fraction of a second out of sync, the object moves into a domain not accessible to our immediate present. The observer sees it vanish because it no longer shares the same present block coordinate. While physically still near the same spatial coordinates, it might be a microsecond forward on its timeline, invisible to sensors locked to the observer's timeline.

## The Role of Exotic Matter in UAP Technology

We discussed in previous chapters that stabilizing wormholes or manipulating time might require exotic matter with negative energy density. If UAPs are test platforms or operational craft using such technology, their propulsion systems likely involve advanced quantum-field engineering.

For example, a UAP might have ring-shaped structures or internal mechanisms that create and maintain regions of negative energy, shaping a controllable bubble of altered spacetime around it. By adjusting this bubble's parameters, the craft could move silently, free from drag and gravitational pull. Observers on the ground would see impossible motion because they assume standard aerodynamics, not realizing the craft isn't "flying" in the conventional sense; it's gliding through a manipulated temporal field.

This kind of technology would remain hidden in classified research facilities. Pilots or military personnel who accidentally encounter these craft might describe them as "beyond next-generation," leaving top brass perplexed. Officials may respond with guarded acknowledgments of strange sightings, knowing they can't reveal the underlying time-based engineering.

## Correlations with Sensitive Locations and Historical Patterns

UAP reports often cluster around military bases, nuclear facilities, or sensitive testing grounds. Traditional explanations suggest reconnaissance by foreign powers or aliens monitoring humanity's technological progress. But if these sites are also where secret time manipulation experiments occur, the correlation makes sense. Covert agencies might test prototypes near secure locations where secrecy can be maintained and quick response teams can retrieve crashed test vehicles or silence witnesses.

If block navigation research began decades ago, perhaps spurred by Cold War desperation, some UAP sightings from the mid-20th century might have been early prototypes. Their erratic flight patterns and occasional malfunctions would explain contradictory witness testimonies. Over time, as the technology improved, sightings might reveal more sophisticated anomalies: craft accelerating from standstill to hypersonic speed instantly, or performing motion patterns not achievable by any known propulsion.

This timeline of improvements hidden behind UAP sightings parallels the evolution of stealth aircraft during the 1970s and 1980s, which were often mistaken for UFOs. The difference here is the technology is even more radical, linked to temporal anomalies instead of mere radar invisibility.

### Could Some UAPs Be Probes from Another Block Coordinate?

Another provocative angle arises if we consider not just local timeline manipulation but full temporal navigation. If future civilizations, or even future branches of our own secret agencies, develop stable block navigation, they might send unmanned probes to another coordinate for observation and data gathering. These probes, designed to minimize paradoxes, could dart in and out of past time coordinates, capturing information about technological progress, climate conditions, or social structures.

From a 21st-century vantage point, such a probe would appear as a UAP: a silent, elusive object with no identifiable propulsion. Its strange behavior, sometimes appearing translucent, other times vanishing instantly, could result from it phasing in and out of our temporal slice. By the time local authorities scramble jets, the probe shifts a fraction of a second forward or backward, escaping detection.

If correct, some UAP sightings might represent "historical surveys" conducted by future observers studying their own past, our present. The notion that we are living in a timeline subject to future scrutiny recasts UAPs not as visitors from distant planets but as tourists from distant centuries.

### Instrumental Data and Physical Evidence

Pilots and radar operators have recorded unusual radar returns from UAPs: blips moving faster than any jet, or stationary objects at high altitude defying wind. How might temporal anomalies appear on instruments?

Radar signals rely on reflecting electromagnetic waves off solid objects. If a craft manipulates spacetime around it, radar pulses might behave strangely. They could bounce back with unusual Doppler shifts, or the craft might appear to "jump" positions between radar sweeps. On infrared sensors, the craft's thermal signature might not align with expected patterns because it's not generating conventional thrust. Instead, it might appear as a cool or anomalously distributed heat source, as exotic matter structures distort energy distribution.

In rare cases, physical evidence like unusual radiation readings or disturbances in local magnetic fields could be found at UAP encounter sites. These anomalies might reflect temporary changes in the vacuum energy density or the presence of field gradients associated with time manipulation attempts. Without understanding temporal physics, investigators would label the evidence as inexplicable.

## Public Perception and the UAP Narrative

The complexity of explaining UAPs as temporal anomalies poses a communication challenge. Most people, steeped in linear notions of time, find it easier to entertain alien spacecraft than vehicles slipping through temporal gradients. Governments, if aware of these truths, may prefer the public to speculate about extraterrestrials or advanced drones rather than the truth: time itself being a frontier of clandestine research.

Popular culture often depicts UFOs as shiny saucers from distant galaxies. Shifting this narrative toward block universe phenomena would require educational leaps. It's simpler to let extraterrestrial stories proliferate, providing cover for secret programs. Meanwhile, scientists pushing quantum gravitational research or wormhole theories might unknowingly develop exotic materials or other theoretical tools that covert projects require, never suspecting that UAP sightings offer real-world hints of applied temporal technology.

If any whistleblowers hint at time manipulation, they risk disbelief. A claimant suggesting UAPs are future probes or local time-bending craft will be dismissed as eccentric. Thus, the status quo remains intact, preserving secrecy and confusion.

## Cultural and Philosophical Implications

If UAPs are linked to temporal anomalies, what does this mean for our understanding of reality? The realization that we are observing manifestations of advanced spacetime manipulation, either from secret human projects or future entities, challenges our self-image as a species confined to one timeline. It suggests a universe richer and more flexible than we imagine.

This perspective resonates with age-old philosophical questions about fate, free will, and destiny. If some UAPs are glimpses of temporal technologies, we might inhabit a world where future knowledge seeps back into the present, subtly guiding or observing our civilization's trajectory. For some, this is empowering, implying we are part of a grand, interconnected continuum of events. For others, it's unsettling, are we lab specimens in a temporal experiment?

Art, literature, and speculative science fiction have occasionally flirted with the idea that UFOs are time travelers. Now, considering the physics of block universe theory and negative energy fields, these once-fanciful ideas gain a veneer of scientific plausibility, reshaping cultural narratives.

**Future Research and Transparency**

Acknowledging a link between UAPs and temporal anomalies would require more than eyewitness accounts. It demands interdisciplinary research, blending aerospace engineering with quantum field theory, gravitational physics, and experimental cosmology. If scientists seriously entertain this idea, they might look for specific signatures: gravitational lensing effects at small scales, unusual temporal correlations in sensor data, or patterns in sightings corresponding to known anomalies in fundamental physics experiments.

Genuine transparency would help. If governments or research institutes ever reach a point where acknowledging these connections becomes beneficial, perhaps when the public can handle the truth or when rivalry among states eases, official briefings might disclose partial information: that some UAP behaviors hint at advanced spacetime engineering, though details remain classified.

Imagine a future conference where aerospace engineers, quantum physicists, and anthropologists discuss UAPs not as extraterrestrial enigmas but as a natural consequence of emerging temporal sciences. This shift would elevate the discourse from tabloid curiosity to a serious scientific frontier.

**Expanding the Discussion**

To further deepen the exploration of UAPs as potential manifestations of temporal anomalies, it is essential to examine historical patterns and technological milestones that could align with reported sightings. This involves cross-referencing UAP incidents with significant events in temporal manipulation research, analyzing potential correlations between technological advancements and the frequency or nature of sightings, and considering the socio-political contexts that might drive the development and deployment of such technologies.

**UAP Origins**

Unidentified aerial phenomena have long occupied a gray area of public perception. Introducing the concept of temporal anomalies and block navigation/manipulation technology into the conversation offers a radically different perspective. Instead of searching distant star systems for the origins of these craft, we might consider that their origin lies in hidden laboratories here on Earth or in human (or post-human) civilizations spanning different eras of our own timeline.

By viewing UAPs through the prism of block navigation/manipulation, many of their inexplicable characteristics become more understandable: their impossible maneuvers are products of altered spacetime geometries, their appearances and disappearances result from temporal phase shifts, and their presence near sensitive installations reflects covert testing or surveillance missions. Even extraterrestrial visitation theories remain possible, but time-based explanations broaden the scope, acknowledging that the universe's complexity extends far beyond simple linear narratives.

In the final analysis, connecting UAPs to temporal anomalies challenges us to think bigger, integrate cutting-edge physics, and question the limits of secrecy and knowledge. The mysteries in our skies may not only be about who is visiting us but also about how time itself can be folded, stretched, and navigated by those with the knowledge and power to do so.

## Historical Patterns and UAP Sightings

Analyzing the timeline of UAP sightings in relation to historical events in advanced physics research could provide insights into possible connections. For instance, increased reports of UAPs might correlate with periods of significant breakthroughs in quantum mechanics, spacetime manipulation theories, or exotic matter research. During the Cold War, when both the United States and the Soviet Union invested heavily in secretive technological advancements, UAP sightings reportedly increased, suggesting a link between military research and aerial anomalies.

Similarly, advancements in stealth technology, propulsion systems, and energy manipulation could directly impact the capabilities of UAPs, making them appear more mysterious and difficult to detect. The development of quantum computing and its potential applications in manipulating spacetime could also play a role in enhancing UAP technologies, allowing for more sophisticated temporal maneuvers that evade conventional detection methods.

## Technological Milestones and UAP Characteristics

Certain technological milestones might align with shifts in the characteristics of UAP sightings. For example, the introduction of more advanced radar systems and infrared sensors could lead to the detection of previously unseen phenomena, altering public perception and increasing reports of UAPs. Conversely, if temporal manipulation technologies enable craft to phase in and out of visibility seamlessly, this could explain the transient nature of many sightings and the lack of physical evidence.

The evolution of propulsion technologies, especially those hypothesized to manipulate spacetime, could also influence UAP behaviors. Enhanced propulsion systems might allow for near-instantaneous acceleration and deceleration, extreme maneuverability, and the ability to hover without apparent means of support. These capabilities would defy current aerospace engineering principles, reinforcing the enigmatic nature of UAPs and fueling speculation about their origins and purposes.

## Socio-Political Context and Temporal Technology Development

The socio-political environment plays a crucial role in driving the development and deployment of advanced technologies, including those related to temporal manipulation. Periods of heightened geopolitical tension, such as the Cold War or the rise of new global powers, often spur significant investments in defense research and covert technological projects. The pursuit of temporal manipulation capabilities could be motivated by the desire to gain strategic superiority, ensure national security, or maintain geopolitical dominance.

Government policies, funding priorities, and international relations significantly influence the direction and pace of technological advancements. In a world where temporal manipulation offers unparalleled strategic advantages, governments might prioritize secrecy, allocate substantial resources to research, and establish specialized agencies or task forces dedicated to exploring these capabilities. This concentration of power and resources could lead to an ultra-secretive arms race, where nations vie to achieve temporal dominance while keeping their advancements hidden from rivals and the public.

**Investigative Approaches to UAP and Temporal Anomalies**

To substantiate the hypothesis that UAPs are manifestations of temporal anomalies or block universe navigation technologies, a multi-disciplinary investigative approach is necessary. This involves combining data from aviation reports, military intelligence, scientific research, and historical records to identify patterns and correlations that support the theory.

**Data Collection and Analysis**

Comprehensive data collection is the first step in investigating UAPs as potential temporal anomalies. This includes gathering eyewitness accounts, radar data, infrared and optical sensor readings, and physical evidence from UAP encounter sites. Advanced data analysis techniques, such as pattern recognition, machine learning, and statistical modeling, can help identify consistent features and anomalies that distinguish UAPs from conventional aircraft or natural phenomena.

Cross-referencing UAP sightings with classified research projects, technological breakthroughs, and historical events can reveal potential links between temporal manipulation research and aerial anomalies. Additionally, analyzing the geographical distribution of sightings, whether they cluster near specific research facilities, military bases, or technologically advanced regions, can provide further evidence supporting the temporal anomaly hypothesis.

**Experimental Research and Simulation**

Engaging in experimental research to simulate the conditions under which temporal anomalies could produce UAP-like behaviors is crucial. This involves developing theoretical models of how spacetime manipulation might manifest in observable phenomena, creating simulations to predict UAP characteristics, and conducting controlled experiments to test these models.

For instance, researchers could simulate how altered spacetime metrics affect light propagation, radar reflections, and thermal signatures. By comparing these simulations with actual UAP data, scientists can assess the plausibility of temporal manipulation as an explanation for the observed anomalies. Collaborations between physicists, aerospace engineers, and data scientists are essential to develop robust models and validate hypotheses.

**Collaboration with Classified Research Units**

Gaining access to classified research units or establishing discreet collaborations with insider sources could provide invaluable insights into the development and deployment of temporal manipulation technologies. Whistleblowers, leaked documents, and insider testimonies might offer concrete evidence of secret projects linked to UAP sightings, helping to substantiate the connection between temporal anomalies and aerial phenomena.

However, such collaborations pose significant ethical and security challenges. Ensuring the protection of sources, maintaining operational security, and navigating the complexities of classified information require careful planning and adherence to ethical standards. Independent oversight and transparent investigative frameworks are essential to balance the pursuit of truth with the protection of sensitive information.

**Addressing Skepticism and Alternative Explanations**

While the temporal anomaly hypothesis offers a compelling explanation for UAPs, it is essential to address skepticism and consider alternative explanations. Conventional theories, such as secret military aircraft, extraterrestrial visitors, atmospheric phenomena, and optical illusions, remain viable explanations for many UAP sightings. Rigorous testing and validation are necessary to differentiate between these possibilities and the temporal anomaly hypothesis.

**Scientific Scrutiny and Peer Review**

Subjecting the temporal anomaly hypothesis to scientific scrutiny and peer review is crucial for its acceptance within the scientific community. Publishing research findings, presenting evidence at conferences, and engaging in open dialogue with fellow scientists can help validate the theory and address criticisms. Independent replication of experiments and verification of data are essential steps in establishing the credibility of the temporal anomaly explanation for UAPs.

## Multi-Disciplinary Validation

Engaging multiple disciplines, including physics, engineering, psychology, and sociology, can provide a holistic understanding of UAP phenomena. Collaboration across these fields can uncover nuanced insights, identify potential biases, and develop comprehensive models that account for both physical anomalies and human perceptions. Multi-disciplinary validation ensures that the hypothesis is robust, resilient, and capable of addressing complex phenomena from various angles.

## Transparent Methodologies

Adopting transparent methodologies in researching UAPs and temporal anomalies fosters trust and credibility. Sharing data, detailing research protocols, and openly discussing findings with the scientific community and the public can demystify the investigation process and mitigate skepticism. Transparency also encourages accountability, ensuring that conclusions are based on evidence and sound reasoning rather than speculation or conjecture.

## Implications for Future Research and Policy

If the temporal anomaly hypothesis gains traction as a valid explanation for UAPs, it would have profound implications for future research and policy. The recognition of temporal manipulation technologies as a reality would necessitate a re-evaluation of scientific theories, technological development priorities, and global governance structures.

## Advancing Temporal Physics

Acceptance of temporal anomalies as a component of UAP phenomena would propel advancements in temporal physics and spacetime manipulation. Research efforts would intensify, focusing on understanding the fundamental principles of temporal navigation, developing stable spacetime structures, and harnessing exotic matter for practical applications. This could lead to breakthroughs in quantum gravity, wormhole stabilization, and the creation of functional block universe navigational systems.

## Ethical and Regulatory Frameworks

The development and deployment of temporal manipulation technologies would require robust ethical and regulatory frameworks to prevent misuse and ensure responsible use. International agreements akin to nuclear non-proliferation treaties might be necessary to govern the distribution and application of temporal technologies. Ethical guidelines would need to address issues such as temporal sovereignty, the rights of individuals across timelines, and the prevention of temporal paradoxes and unintended consequences.

## National and Global Security

Temporal manipulation capabilities would redefine national and global security dynamics. Nations possessing such technologies would hold unprecedented strategic advantages, necessitating new forms of deterrence and defense. Global security policies would need to adapt to address the unique challenges posed by temporal warfare, espionage, and surveillance. Diplomatic efforts would be essential to establish trust, prevent temporal conflicts, and promote the peaceful use of temporal technologies.

## Integrating Temporal Anomalies into UAP Research Programs

To effectively incorporate the temporal anomaly hypothesis into UAP research programs, a structured and systematic approach is required. This involves establishing dedicated research units, securing funding for interdisciplinary studies, and fostering collaborations between government agencies, academic institutions, and independent researchers.

### Establishing Dedicated Research Units

Governments and international bodies should consider establishing dedicated research units focused on exploring the temporal anomaly hypothesis within UAP studies. These units would coordinate data collection, conduct experimental research, and develop theoretical models to understand the temporal aspects of UAP phenomena. By centralizing efforts, these units can ensure a comprehensive and unified approach to investigating the temporal dimensions of UAPs.

### Securing Funding for Interdisciplinary Studies

Temporal anomaly research requires significant funding to support interdisciplinary studies that span physics, engineering, computer science, and data analysis. Securing funding from government grants, private institutions, and international organizations can facilitate the development of advanced research facilities, acquisition of cutting-edge technology, and recruitment of top-tier scientists and engineers.

### Fostering Collaborations Between Agencies and Academia

Collaboration between government agencies, military research units, and academic institutions is essential for advancing the temporal anomaly hypothesis. Joint research projects, shared resources, and knowledge exchange programs can enhance the quality and scope of investigations. Encouraging open communication and collaboration fosters innovation and accelerates the discovery of new insights into UAP phenomena.

## The Future of UAP Research and Temporal Anomalies

As our understanding of physics and technology continues to evolve, the exploration of temporal anomalies as an explanation for UAPs remains a frontier of scientific inquiry. Continued research, interdisciplinary collaboration, and open-minded exploration are essential for unraveling the mysteries of unidentified aerial phenomena and their potential connections to block universe navigation.

## Embracing the Unknown

The pursuit of understanding UAPs through the lens of temporal anomalies embodies the spirit of scientific exploration, embracing the unknown, challenging existing paradigms, and seeking answers to the most profound questions about our universe. It encourages scientists and researchers to push the boundaries of knowledge, explore new theoretical frameworks, and develop innovative technologies that could reshape our understanding of reality.

## Preparing for Paradigm Shifts

The acceptance of temporal anomalies as a component of UAP phenomena would signify a paradigm shift in physics and aerospace engineering. Preparing for such shifts involves fostering a culture of scientific curiosity, encouraging unconventional thinking, and supporting bold research initiatives that dare to explore the fringes of known science. Educational institutions, research organizations, and policy makers must be ready to adapt to new discoveries and integrate them into the broader fabric of scientific knowledge.

**Long-Term Implications for Humanity**

The implications of discovering temporal manipulation technologies extend beyond scientific and technological advancements, they touch the very essence of human existence. Understanding and harnessing the power to navigate the block universe could lead to profound changes in how we perceive time, history, and our place in the cosmos. It could enable humanity to avert disasters, explore alternate timelines, and gain unprecedented insights into the nature of reality.

However, with great power comes great responsibility. The ethical and moral considerations discussed earlier would become even more critical as humanity gains the capability to influence timelines and shape the course of history. Balancing technological advancement with ethical stewardship would be essential to ensure that such power is used for the collective good, preserving the integrity of the temporal fabric and safeguarding the future of civilization.

**Final Reflection**

Unidentified aerial phenomena, long a source of intrigue and speculation, may hold deeper connections to the fundamental structure of spacetime and the block universe theory. By considering UAPs as potential manifestations of temporal anomalies, we open new avenues for understanding these enigmatic sightings and the advanced technologies that might underpin them.

This exploration not only enriches our comprehension of UAPs but also challenges us to rethink the boundaries of human knowledge and technological capability. It compels us to confront ethical dilemmas, redefine security paradigms, and embrace the mysteries that lie at the intersection of time, space, and human ingenuity.

In the quest to unravel the secrets of the skies, the temporal anomaly hypothesis stands as a testament to the ever-expanding horizons of scientific inquiry and the enduring human spirit of exploration. Whether UAPs are relics of advanced human experiments, echoes of future civilizations, or manifestations of extraterrestrial intelligence, their study propels us toward a deeper understanding of the universe and our place within it.

# Chapter 14

## Integrating Consciousness Studies and Spacetime

A crucial element remains largely uncharted: consciousness itself. Consciousness is the lens through which reality is experienced, including the flow of time. Without sentient observers, concepts like past, present, and future lose much of their significance. It's our inner awareness that transforms a static four-dimensional universe into a lived narrative, complete with memories, anticipations, and interpretations.

This chapter explores how recent advancements in consciousness studies can enrich our understanding of spacetime. It examines how the subjective experience of time might mirror deeper structures within the brain, considers how quantum or relativistic theories of mind could intersect with block universe coordinates, and investigates how emerging fields like neurophenomenology and integrated information theory might bridge these complex ideas. The goal is to highlight that fully comprehending time may require grappling with the fundamental nature of consciousness, a task that benefits from interdisciplinary dialogue across physics, neuroscience, cognitive science, philosophy, and beyond.

### The Puzzle of Consciousness in a Block Universe

If the block universe view holds true, all events, past, present, and future, exist simultaneously. Yet, human consciousness perceives them as unfolding in a linear sequence. This discrepancy presents a profound puzzle: Why do we experience the flow of time so vividly if, fundamentally, there might be no actual "flow"?

One possible explanation lies in the nature of consciousness itself. Imagine watching a flipbook, each page static, yet when flipped rapidly, the images appear to move. Similarly, consciousness could be stitching together these static moments into a coherent, flowing experience. It's as if the block universe contains all spacetime frames at once, while consciousness "plays" them in sequence. This creates the vivid sensation of time passing, even if, on a grand scale, all events coexist without temporal progression.

This perspective invites collaboration between physicists and consciousness researchers. Physicists might ask: How does the brain's temporal processing relate to spacetime geometry? Consciousness scholars might ponder: Does the mind's temporal structure reflect underlying causal constraints in the universe, or is it simply a mental construct evolved for survival?

**Neural Correlates of Temporal Experience**

Neuroscience has made significant strides in identifying the neural correlates of temporal perception. Certain brain regions, such as the basal ganglia and cerebellum, are involved in timing intervals, while the hippocampus and related structures store episodic memories that give a sense of personal history. The prefrontal cortex, on the other hand, helps plan future actions, effectively simulating what is yet to come.

Consider the way memories of past events can transport someone back to a moment long gone, like reliving the details of a favorite book's plot. From a block universe standpoint, these neural mechanisms "slice" the four-dimensional manifold into a linear experience, selecting which "now" is felt and how it connects to remembered "thens" and anticipated "whens." Brain imaging studies reveal that when recalling a past event, the brain activates similar neural patterns to those formed during the original experience. This neural replay gives a sensation of traveling back in time, albeit purely subjectively.

Viewing the brain as a sophisticated spacetime interpreter suggests that neural temporal processing mirrors the complexity of a block universe. The mind imposes order on a static reality, carving out a dynamic narrative. Understanding this translation from a timeless block to a time-rich experience could unveil principles applicable to both consciousness and cosmology.

## Quantum Theories of Consciousness and Spacetime

Some speculative theories propose links between quantum mechanics and consciousness. While mainstream neuroscience doesn't require quantum effects to explain cognition, a few researchers speculate that quantum states in microtubules or other neural substrates might play a role. If future research supports quantum contributions to consciousness, how might this intersect with spacetime?

Quantum mechanics and relativity are notoriously difficult to reconcile. Yet, if consciousness is quantum-linked, it could serve as a novel bridge. For instance, if conscious observation "collapses" quantum wavefunctions, might it also influence how time is perceived, collapsing a probabilistic spread of future events into a definite sequence?

Though these ideas remain highly speculative, they open imaginative avenues. Perhaps the subjective arrow of time partly emerges from the interplay between conscious observation and quantum uncertainty. As consciousness "selects" outcomes, it generates the impression of temporal progression. In this view, consciousness wouldn't alter the block universe but would shape which paths become salient in an observer's timeline, providing a psychological ordering to events that are, at their core, static.

## Integrated Information Theory and Temporal Structure

Integrated Information Theory (IIT), a prominent framework in consciousness studies, posits that consciousness corresponds to the integration of information within a system. According to IIT, the level of consciousness (phi) relates to how much the system's structure and causal interactions cannot be reduced to independent parts. This theory emphasizes the complexity and interconnectedness of neural activity.

How could IIT inform our understanding of time in a block universe? If consciousness arises from integrated information, then temporal integration, how the brain binds successive moments into a coherent narrative, might be crucial. The brain's ability to form causal links between past sensory inputs and present states could create a dynamic pattern of integrated information extending across time.

Imagine a chef who is meticulously following a recipe. Each step builds on the previous one, creating a seamless dish. Similarly, the brain integrates information across time, crafting a coherent experience from discrete moments. In a block universe, every slice of time exists independently. However, a conscious agent integrates these slices, at least those it experiences, into a causal tapestry. IIT might suggest that consciousness endows certain worldlines with meaningful structure. The block universe is like a four-dimensional library; consciousness picks a volume and reads it linearly, forging meaning by integrating events into a narrative. Understanding how integrated information spans temporal intervals could provide insights into why a singular, flowing "now" is experienced.

## Neurophenomenology: Linking Subjective Time and Brain Dynamics

Neurophenomenology, pioneered by Francisco Varela, seeks to bridge the gap between subjective experience (phenomenology) and objective brain data (neuroscience). By collecting first-person reports of temporal experience, how duration is sensed, futures are anticipated, or pasts are recollected, alongside neural recordings, researchers aim to identify patterns that correlate subjective time with measurable brain states.

If the block universe is the ultimate objective backdrop, an unchanging spacetime structure, then subjective temporal experience is the "user interface" that conscious observers employ. Neurophenomenology can serve as a methodology to decode this interface. Through carefully designed experiments, participants might describe how their sense of time shifts during meditation, hypnosis, or virtual reality simulations. Simultaneously, EEG, fMRI, or MEG recordings track brain oscillations, connectivity patterns, and activity gradients over time.

Correlating these data sets might reveal that certain neural configurations produce a heightened sense of temporal continuity, while others fragment time into disjointed snapshots. This would not only enrich consciousness studies but also offer clues about how brains construct temporal narratives from a timeless reality, insights potentially valuable to physicists modeling observer-dependent temporal slicing.

## The Emergence of a Subjective Arrow of Time

Physicists often attribute the arrow of time to thermodynamics: entropy increases from past to future. Yet, entropy alone doesn't explain why time is experienced as moving forward. Consciousness may play a role in accentuating this arrow. After all, memories are of the past, not the future. Could the asymmetry in memory formation be tied to how consciousness integrates information?

If memory encoding relies on irreversible biochemical changes, like strengthening certain synapses, then our cognitive architecture inherently imposes a temporal direction. We store traces of what came before, not what lies ahead, imparting a subjective arrow of time. Consciousness, anchored in these asymmetric memory processes, experiences a world moving from a known past to an unknown future.

From this perspective, the subjective arrow of time emerges at the intersection of physical laws and cognitive architecture. While the block universe contains all events equally, consciousness is built on memory systems that only reference earlier states, never later ones. Integrating consciousness studies and spacetime thus clarifies that our time sense is not purely physical or purely mental but a joint product of both domains.

## Consciousness as a "Focal Point" in the Block Universe

Imagine the block universe as a grand four-dimensional tapestry. In this view, consciousness might resemble a spotlight that illuminates one thread (the observer's worldline) and one "point" (the present moment) at a time. But unlike a static spotlight, consciousness moves along this thread, weaving meaning from a succession of illuminated events.

Incorporating ideas from consciousness research, such as how attention focuses on certain stimuli and filters out others, could inform physicists developing models of observer-based measurements. Physicists have long recognized that measurement and observation play key roles in quantum mechanics. Perhaps consciousness, with its selective attention, acts as a similar filter at the classical scale, highlighting one temporal slice after another.

This integration suggests a new metaphor: the block universe is a gallery of paintings (events), and consciousness is the curator who arranges them into a coherent exhibit (experience). Studying how attention and cognition structure the perception of time may help unify how observers are conceptualized in both physics and philosophy.

**Potential Experiments and Interdisciplinary Collaborations**

Practical investigation of these ideas requires an interdisciplinary approach. Organizing workshops that bring together physicists, neuroscientists, and philosophers can facilitate the design of experiments testing hypotheses about the mind's temporal structure. For instance, can altering the brain's integration windows, using neural stimulation to change how sensory input is consolidated, affect subjective time perception in ways predicted by certain spacetime models?

Neuroscientists might develop brain-machine interfaces that feed back altered temporal signals, such as artificially slowing down or speeding up sensory updates, to observe how consciousness of time flow is reshaped. Physicists could contribute by constructing analogies that mirror how relativistic frames perceive simultaneity differently, exploring whether manipulating sensory latencies can mimic these relativistic effects in subjective experience.

Philosophers and cognitive scientists could refine conceptual frameworks, ensuring that terms like "temporal experience," "block universe," and "observer" are used consistently across fields. The goal isn't to prove metaphysical claims but to open fertile territory for novel research, where insights into cognition guide theoretical physics and vice versa.

**Ethical and Existential Implications**

Examining consciousness and spacetime together raises profound existential questions. If all times exist simultaneously, yet consciousness experiences them linearly, what does this mean for free will? Are actions merely "read" from a predetermined script, or is consciousness actively shaping the narrative perceived?

Moreover, as research into neural time manipulation advances, such as using virtual reality to distort the sense of duration, ethical concerns arise. Could environments be created that intensify the subjective arrow of time, trapping individuals in a prolonged "now"? Or, conversely, might therapeutic interventions help trauma survivors revisit and reframe painful past events, altering their emotional resonance?

Understanding how consciousness interprets and navigates time offers tools with both liberating and potentially manipulative applications. Integrating consciousness studies and spacetime is not just academically intriguing; it challenges the responsible handling of new insights into subjective reality.

## Conclusion

Integrating consciousness studies and spacetime offers a synthesis that deepens the understanding of both domains. Recognizing that consciousness shapes the experience of time and that the block universe demands new perspectives on observers creates common ground for collaboration among neuroscience, philosophy, and physics.

The linear flow of time may be, at least in part, a construct of conscious cognition, a way for the mind to navigate a timeless landscape. Memory, attention, and sensory integration create the impression of temporal progression. Advanced theoretical frameworks, like Integrated Information Theory or neurophenomenology, help map the neural underpinnings of this process, linking subjective reports to brain dynamics.

While no single theory currently unifies these fields, the direction is clear, to truly understand time, the role of consciousness must be considered, and vice versa. As quantum theories are refined, relativistic effects are explored, and richer models of cognition are developed, the picture becomes more holistic. In that picture, the block universe's silent grandeur finds expression in the mind's narrative weaving, and the elusive riddle of consciousness gains new dimensions by embracing the full breadth of spacetime.

This integration affirms that no part of nature, neither the tapestry of events nor the minds that observe it, stands alone. Just as light needs an observer to be seen, the block universe's meaning emerges only when filtered through conscious experience. Moving forward requires forging alliances between disciplines, aiming for insights that honor both the complexity of spacetime and the mystery of consciousness.

## Chapter 15

## The Simulation Hypothesis and the Block Universe

In previous chapters, we explored how the block universe hypothesis can recasts our understanding of time, presenting it as a dimension that coexists with the three spatial dimensions rather than something that flows. This perspective likens the block universe to a four-dimensional tapestry, containing all events, past, present, and future, as a static structure. However, a lingering question arises: if all times exist equally and our linear experience of time is just one perspective, could the entire structure of spacetime resemble a stored, pre-written "program" running on an advanced computational substrate?

This inquiry leads us to the simulation hypothesis, which suggests that our reality might be a meticulously engineered simulation executed by an extremely advanced intelligence or computational system (Bostrom, 2003). Linking the block universe to the notion that we may exist inside an advanced "matrix" (Chalmers, 2005) is speculative and contentious, yet it opens intriguing conceptual doors. If spacetime is envisioned as data held and processed by a higher-level computational entity, the block universe might not only reflect a static geometric reality but also represent a stable, unchanging dataset within a colossal computational architecture.

This chapter explores how the simulation hypothesis interacts with the block universe concept. It discusses how leading thinkers have approached the idea of simulated reality, considers what it means for spacetime to be computable, and examines whether the block universe view supports the plausibility or conceptual coherence of living in something akin to a matrix. Although highly speculative, this exploration may help us understand our roles as observers and conscious entities embedded in a structured reality that could be fundamentally digital.

## The Simulation Hypothesis: A Brief Overview

Philosopher Nick Bostrom is perhaps the most cited advocate of the simulation hypothesis. In his 2003 paper "Are You Living in a Computer Simulation?" (Bostrom, 2003), he argues that one of three propositions is likely true: (1) advanced civilizations never reach a stage capable of running simulations of their ancestors, (2) they reach that stage but choose not to run such simulations, or (3) we are almost certainly living in a simulation. Bostrom's argument is philosophical and probabilistic rather than empirical, but it has sparked widespread discussion. His logic hinges on the idea that if civilizations can run many ancestor simulations, the odds that we are in one might be overwhelming.

Since Bostrom's initial proposition, others have contributed to the debate. Rizwan Virk (2019) in "The Simulation Hypothesis" connects the dots between quantum physics, information theory, and the idea of a simulated universe, while philosopher David Chalmers (2005) treats the matrix-like scenario as a metaphysical proposition: a simulated reality is not less "real" but simply operates on a different substrate than we imagine. In the scientific domain, physicists like James Gates have found peculiar "error-correcting codes" embedded in the fundamental equations of supersymmetry (Greene, 2011), prompting speculative connections to digital computation.

What unites these thinkers is the premise that reality could be informational at its core. If so, spacetime might be a computational digital construct, a grid of data points processed by an advanced computer. In this view, the block universe, encompassing the entire timeline, resembles a pre-loaded database accessible simultaneously to the system running it.

The notion that the universe might be fundamentally digital is not new. Digital physics, championed by researchers like Edward Fredkin (Fredkin, 2003) and Gerard 't Hooft (Hooft, 2016), proposes that the cosmos can be described as a cellular automaton or a discrete computational network. If the universe is indeed digital, it may be representable by finite data structures and computable algorithms.

In a digital cosmos, time might not "flow" but instead be indexed like frames in a digital simulation. From the perspective of a hypothetical programmer, each moment in the block universe is a saved state. The laws of physics are the code that governs how these states relate to each other. Consciousness, as previously suggested, could be the process running on these states, giving the illusion of time passing by sequentially reading them in order.

**The Block Universe as a Computational Dataset**

Visualizing the block universe within an advanced simulation involves imagining a massive computational array storing the coordinates of every particle, every field configuration, and every event throughout the entire history of the universe, from the Big Bang to the distant future. Each "time slice" is just another index in this array, no more privileged than any other. The "now" we experience corresponds to the simulation's rendering of a particular array index into the conscious minds of simulated observers.

In such a scenario, relativity's relativity of simultaneity might reflect different ways of reading the underlying data structure. Observers moving at different velocities select different "slices" of the block. In computational terms, they apply different sorting criteria or indexing schemes to the same dataset. This consistency with the mathematical constraints of relativity ensures that the simulation aligns with our observed physical laws.

Our subjective experience of time flow and the arrow of time might emerge because the simulation's update cycle enforces a strict reading order for each observer's memory subsystem. Memories of previous frames (past states) are stored, but future frames remain unrendered, not because they do not exist in the dataset, but because the memory functions only add new frames as "past" and never preload future frames. This built-in asymmetry could be a deliberate design choice in the simulation's architecture.

**Indirect Evidence and Philosophical Plausibility**

No direct experiment currently proves we are in a simulation. However, some scientists and philosophers find indirect hints intriguing. For example, the discovery of fundamental constants that seem finely tuned to allow complexity and life (Schellekens, 2013) might be interpreted as settings chosen by simulators. The block universe could serve as the "source code repository," containing all versions and branches of reality.

Skeptics argue that complexity does not equal design. The block universe can arise from elegant mathematical laws without any overarching programmer. Still, the simulation hypothesis resonates with the block universe in one crucial way: both diminish the primacy of the "now." In a simulation, the "now" is just a rendered moment, while all other moments may reside in memory. Similarly, in the block universe, the "now" is an observer-dependent slice through a timeless manifold. Both views de-center human intuition about time's flow and highlight that what feels dynamic could be static from another vantage point.

**Consciousness as a Rendering Mechanism**

Extending the simulation analogy further: in a sophisticated virtual reality system, the rendering engine only displays what the user's perspective requires at any given moment. High-detail graphics are generated on the fly for the user's field of view, while other parts of the game world exist only as potential data, ready to be rendered if the user moves there.

Could consciousness serve a similar function in the block universe simulation? Perhaps consciousness "renders" the slice of spacetime the observer inhabits, accessing the underlying block data as needed. This rendering might be what collapses a vast static dataset into a lived narrative, just as a gamer's experience collapses the entire game map into a sequence of traversed environments.

This notion parallels certain interpretations in quantum mechanics, where observation affects outcomes (Wheeler, 1990). From a simulation perspective, the observer's conscious process might determine how pre-written events become experiential reality. For instance, the simulation might have rules that allocate computational resources efficiently, only "resolving" details of events that conscious observers focus on, much like a game engine optimizes rendering to avoid unnecessary computations.

## Technological Trajectories and Potential Tests

If we are in a simulation, could advanced civilizations running it leave detectable traces, like digital watermarks or quantization artifacts in the fabric of spacetime? Some researchers have proposed searching for limits to the energy of cosmic rays or irregularities in the cosmic microwave background that might indicate a discrete underlying lattice (Silas R. Beane et al., 2012). While speculative, such tests highlight that the simulation idea has inspired attempts to find empirical signatures.

If future experiments reveal anomalies consistent with discretized spacetime, error-correcting codes in nature's laws (Gates, 2011), or boundary conditions reminiscent of a computational environment, that would bolster the simulation hypothesis. The block universe model would fit neatly into such a framework, serving as a perfect dataset stable and eternal, a hallmark of a system stored in some advanced computational substrate.

However, it is essential to note that the absence of evidence or negative results would not disprove the simulation scenario. It could simply mean that the simulators are exceptionally good at maintaining a seamless illusion or that the universe's computational substrate is beyond our current detection methods.

## Ethical and Existential Implications

Much like the block universe challenges our notions of free will and determinism, the simulation hypothesis adds another layer of complexity. If we live in a simulation, are our choices "real," or are they programmed outcomes of complex algorithms? The block universe, already diminishing the importance of a flowing present, combined with the simulation argument, suggests we might be characters in a grand computational narrative.

Some might find solace in this view: our existence could be meaningful if the simulators have benevolent purposes, like preserving knowledge or ensuring diverse life forms thrive. Others might feel unsettled, questioning the authenticity of their experiences. Practically speaking, whether we are simulated or not changes little about our day-to-day ethics and aspirations. We still strive for meaning, love, and knowledge, much like a character in a well-crafted story can have authentic emotions and moral dilemmas.

## Integrating the Simulation Hypothesis into Our Conceptual Framework

Introducing the simulation hypothesis adds a "who" and "why" dimension to the block universe structure. Until now, the block universe was accepted as a given: a timeless cosmic stage set by impersonal laws. The simulation perspective suggests that advanced intelligences, possibly akin to future versions of ourselves, could be behind the scenes, programming the laws and initial conditions, and perhaps even curating the tapestry of events.

This interpretation reimagines the block universe as a stable data file in a cosmic supercomputer's memory banks. Our arrow of time is then a playback function. The complexity of life, consciousness, and culture emerges like the intricate patterns in a simulated game world, spontaneous yet ultimately contained within algorithmic constraints.

While this remains unproven, it resonates with recent philosophical and scientific discourses, prompting interdisciplinary dialogues. Physicists consider digital ontologies, neuroscientists ponder if the brain's computations could be mimicked perfectly, and philosophers debate whether simulation equates to illusion or a novel form of authentic reality.

### The Block Universe and Computational Physics

The intersection of the block universe and computational physics offers a fertile ground for theoretical exploration. Computational physics posits that the universe operates like a vast computational machine, processing information through fundamental laws akin to algorithms. If the block universe is viewed as a computational dataset, it aligns with the idea that spacetime can be modeled and simulated with sufficient computational power.

In this framework, the block universe is not just a passive backdrop but an active computational process. Each event, each particle interaction, is a data point processed by the underlying computational laws. Consciousness, in this view, might be an emergent property arising from complex information processing within this system, contributing to the dynamic narrative that observers experience.

### Practical Implications and Philosophical Questions

The simulation hypothesis, when combined with the block universe concept, raises several practical and philosophical questions:

**Determinism vs. Free Will**: If reality is a simulation with pre-written events, how much control do individuals truly have over their actions? Does free will exist, or are choices predetermined by the simulation's code?

**Purpose and Meaning**: What is the purpose of the simulation? Are there objectives or goals set by the simulators, and how do these influence the simulated beings' lives and development?

**Knowledge and Discovery**: How does the potential existence of a simulation impact our pursuit of knowledge? Would discovering the simulation's parameters alter scientific research and our understanding of the universe?

**Ethical Responsibilities**: If creators are running simulations, what ethical responsibilities do they hold towards the simulated beings? Conversely, what responsibilities do simulated beings have towards their creators?

### Bridging the Gap Between Theory and Experience

Bridging the gap between the simulation hypothesis and the block universe requires a multidisciplinary approach. Combining insights from physics, computer science, neuroscience, and philosophy can help construct a more cohesive understanding of how these concepts interplay. For instance, advancements in quantum computing and artificial intelligence may provide new tools for modeling and testing aspects of the simulation hypothesis within the block universe framework.

Moreover, exploring the subjective experiences of consciousness through neuroscience and cognitive science can offer clues about how a simulated reality might be experienced by its conscious observers. Understanding how the brain constructs a coherent narrative from a vast, unchanging dataset could illuminate how simulated beings perceive and interact with their environment.

## Potential Challenges and Criticisms

Both the simulation hypothesis and the block universe face significant challenges and criticisms:

**Lack of Empirical Evidence**: Neither hypothesis currently has direct empirical support. The simulation hypothesis relies heavily on probabilistic reasoning, while the block universe is primarily a theoretical construct derived from relativity.

**Falsifiability**: The simulation hypothesis is often criticized for its lack of falsifiability. If simulators are sufficiently advanced, they could potentially hide all evidence of the simulation, making it impossible to test scientifically.

**Philosophical Implications**: The simulation hypothesis raises deep philosophical questions about existence, reality, and consciousness. It blurs the lines between the observer and the observed, challenging traditional notions of objective reality.

**Technological Limitations**: Current technology does not allow us to test the simulation hypothesis directly. While theoretical models and indirect tests offer some avenues, the computational power required to simulate an entire universe is far beyond our current capabilities.

## Future Directions and Research Opportunities

Despite these challenges, the intersection of the simulation hypothesis and the block universe concept presents numerous research opportunities:

**Computational Modeling**: Developing advanced computational models that simulate block universe dynamics within a simulated framework can provide insights into how such a system might operate.

**Quantum Information Theory**: Exploring the connections between quantum information theory and the simulation hypothesis could reveal underlying principles that govern the block universe's computational structure.

**Neuroscientific Studies**: Investigating how the brain processes information and constructs temporal narratives can offer clues about how consciousness might function within a simulated block universe.

**Philosophical Inquiry**: Engaging in philosophical debates about the nature of reality, consciousness, and existence can help clarify the implications of these hypotheses and guide scientific exploration.

**Conclusion: Embracing the Possibility**

Merging the block universe hypothesis with the simulation scenario stretches our imaginations and challenges our understanding of reality. We transition from viewing spacetime as a static geometric entity to envisioning it as a dataset within a cosmic program. This leap, while radical, aligns with the spirit of inquiry that guided earlier chapters, questioning linear time, exploring subjective temporal navigation, and acknowledging that consciousness shapes how events are perceived.

If the block universe is the ultimate data structure and we are simulated observers, then everything we cherish, our history, science, art, emotions, might be lines of code executed by a computational intelligence beyond our comprehension. From this vantage, the grandeur of the block universe and the intricacy of conscious experience blend into a tapestry that is both scientific and mythic: characters in a cosmic software, witnessing time as a stable, eternal array that is played back one moment at a time.

Whether this speculation points to genuine truths or remains an intellectual exercise is unknown. However, asking these questions broadens our conceptual horizons. It prompts us to view the block universe not just as an abstract theory but as a clue that reality itself could be more layered, purposeful, and intricately constructed than previously imagined. By entertaining the simulation hypothesis, scientific rigor is not abandoned; instead, it is pushed to consider new paradigms, ensuring that exploration into the mysteries of time continues with fresh perspectives and renewed curiosity.

# Chapter 16

## The Road Ahead: Future Research Directions

The exploration of time, space, and consciousness has led to a myriad of fascinating ideas and theories. From the theoretical elegance of the block universe to the formidable engineering required for physical time manipulation, this research has been anything but ordinary. As humankind stands at this crossroads of knowledge, it becomes evident that our current understanding is merely the beginning. More questions have emerged than answers, and numerous paths beckon for future exploration.

This chapter looks at promising research frontiers that hold the potential to deepen our insights into the block universe spacetime and our relationship with it. It spans multiple disciplines, including physics, cognitive science, neuroscience, philosophy, history, ethics, engineering, and even the arts. Achieving meaningful progress will require unprecedented collaboration and creativity. I'll address new technologies, theoretical breakthroughs, experimental strategies, and interdisciplinary dialogues that could help refine our concepts, test our speculations, and perhaps bring us closer to a reality where navigating the block universe becomes more tangible.

### Advancing Theoretical Physics and Quantum Gravity

At the heart of many unresolved questions lies the challenge of unifying general relativity, which underpins the block universe concept, with quantum mechanics. A quantum theory of gravity remains elusive, yet it is essential for understanding whether spacetime is fundamentally discrete or continuous, how wormholes might form or stabilize, and whether negative energy and exotic matter can exist in stable forms.

Future research in string theory, loop quantum gravity, or other quantum gravity frameworks may illuminate whether block universe navigation is theoretically possible. Imagine physicists poring over complex equations, hoping to discover a stable wormhole solution that doesn't require unrealistic conditions. If new mathematical solutions indicate that stable wormholes or closed timelike curves could exist under specific energy configurations, it would bolster the idea that advanced engineering might one day realize them. On the flip side, if theories reveal that such phenomena demand impossibly high energies, some of the more fanciful ideas about time travel might be set aside.

Another intriguing avenue is refining our understanding of the "observer" in physics. Quantum mechanics already suggests that measurement and observation play special roles. Integrating consciousness or observer-dependent frameworks could lead to groundbreaking insights. For instance, researchers might explore how the brain's temporal processing relates to spacetime geometry, blurring the lines between physics and cognitive science. This could inspire joint research projects where physicists and cognitive scientists model the observer as a dynamic entity shaping perceived temporal structure.

Moreover, exploring the implications of the holographic principle, which posits that all the information contained within a volume of space can be represented as encoded data on the boundary of that space, could offer new perspectives on the nature of spacetime. If spacetime itself is emergent from more fundamental informational structures, understanding these underlying layers could provide the keys to manipulating time and space in ways previously thought impossible.

**Experimental Approaches to Temporal Geometry**

Testing block universe predictions poses significant challenges, primarily because the block universe posits that all of time is equally real. However, advancements in experimental techniques may open new doors. Precision instruments like gravitational wave detectors, atomic clocks, and quantum sensors are continually improving and may one day detect minute spacetime fluctuations indicative of exotic phenomena such as negative energy densities or proto-wormholes.

Consider the possibility of advanced tabletop experiments or high-energy particle colliders creating conditions that mimic aspects of curved spacetime. Scientists could study how quantum fields behave under extreme temporal geometries, much like how analogy models using Bose-Einstein condensates mimic black holes. While simulating a wormhole in a laboratory remains a distant dream, extending these analogies to temporal structures might provide valuable insights.

Moreover, as quantum computing matures, quantum simulators could model exotic spacetime configurations. Programmable quantum devices might serve as "toy universes," allowing researchers to probe theoretical predictions and gain intuition about time manipulation indirectly. Imagine a team of scientists using a quantum simulator to test how different energy distributions affect spacetime curvature, inching closer to understanding the feasibility of temporal shortcuts.

In addition to direct experimentation, developing sophisticated computational models that simulate the block universe under various conditions could help identify potential signatures of time manipulation. These models could integrate principles from quantum gravity, information theory, and complex systems to predict how spacetime might behave under different theoretical frameworks. By comparing these predictions with observational data, scientists could validate or refute existing theories, paving the way for new discoveries.

**Neuroscience and Cognitive Science of Temporal Experience**

On the human side, understanding how the brain constructs the sense of flow, sequence, and the arrow of time is crucial. Integrative studies combining brain imaging, behavioral experiments, and computational models are needed to unravel the neural mechanisms of temporal perception. For example, researchers could manipulate sensory delays or feedback loops to observe how subjects adapt to altered temporal cues. Virtual reality environments might create illusions of sped-up or slowed-down events, testing whether and how these manipulations change neural activity patterns.

Such studies could map subjective reports onto observed brain dynamics, refining theories about which neural circuits give rise to the feeling of "now," memory recollection, and anticipation. Imagine participants in a lab donning VR headsets that distort their perception of time, while neuroscientists monitor their brain activity. These experiments could reveal how the brain integrates information across time, offering clues about how consciousness shapes our temporal experience.

The implications extend beyond pure research. Understanding the cognitive and neural bases of temporal perception might lead to cognitive tools that improve memory, treat trauma, or alleviate temporal disorientation in neurological conditions. Philosophically, grasping why our brains perceive time linearly could inform debates over free will, determinism, and the nature of meaningful action in a possibly fixed future.

Furthermore, exploring the role of neuroplasticity in temporal perception could uncover how flexible our experience of time truly is. If the brain can adapt its temporal processing mechanisms, it might be possible to train individuals to experience time differently, potentially opening avenues for cognitive enhancements or therapeutic interventions.

**Interdisciplinary Dialogues and Philosophical Engagement**

No single discipline can tackle these challenges alone. Future research directions must encourage philosophers, ethicists, and historians to engage with physicists, neuroscientists, and engineers. Philosophers can clarify conceptual pitfalls, ensuring terms like "observer," "navigation," and "block universe" have consistent meanings. Ethicists can anticipate moral dilemmas that may arise if temporal technologies ever mature, addressing questions like how to prevent abuse or ensure equitable access.

Historians and anthropologists might study how past societies conceptualized time, exploring whether their narratives, myths, and metaphors resonate with modern block universe ideas. Their insights could reveal whether certain conceptual leaps are culturally influenced and how global perspectives differ in handling time's mysteries. By blending these perspectives, researchers can create a richer intellectual environment, increasing the chances of breakthroughs that respect human values and cultural diversity.

For instance, a philosopher might collaborate with a physicist to dissect the concept of simultaneity in relativity, while an ethicist works with engineers to develop guidelines for responsible experimentation with time manipulation technologies. These interdisciplinary collaborations ensure that research is not only scientifically sound but also ethically grounded and culturally sensitive.

Additionally, integrating insights from the humanities can provide a more holistic understanding of how humans relate to time. Literature, art, and philosophy offer unique perspectives on temporal experience, often highlighting aspects that pure science might overlook. Engaging with these fields can inspire new ways of thinking about spacetime and consciousness, fostering innovative approaches that bridge the gap between empirical evidence and human experience.

**Technological Innovations and Simulation Tools**

As computational power grows, sophisticated simulations of hypothetical spacetime configurations become possible. Imagine a research consortium developing a comprehensive "time manipulation simulator" where scientists input theoretical parameters, energy distributions, exotic matter conditions, and watch how simulated universes unfold. Such virtual experiments could highlight stable solutions or expose insurmountable paradoxes, providing a testing ground for theoretical models.

Advances in materials science and energy generation could also inch us closer to manipulating quantum vacuum states or creating tiny negative energy regions. While still speculative, incremental progress in controlling quantum fields, zero-point energy, or advanced metamaterials could, in the future, lay the groundwork for proto-technologies that test the limits of spacetime engineering.

On the cognitive side, augmented and virtual reality interfaces could help researchers and educators visualize complex four-dimensional structures. By immersing themselves in dynamic spacetime simulations, scientists might gain intuitive insights, formulating hypotheses that static equations cannot inspire. These tools democratize complex ideas, allowing even lay audiences to understand why the block universe is so conceptually challenging.

Furthermore, developing artificial intelligence systems capable of modeling and predicting spacetime behavior could accelerate research. AI-driven simulations might uncover patterns or solutions that human researchers might overlook, providing new avenues for exploration and understanding.

**Cultural, Artistic, and Educational Ventures**

Future research directions need not remain confined to laboratories and academic institutions. Artists, writers, and filmmakers can interpret and communicate these concepts, making them accessible and emotionally resonant. Art installations that collapse past, present, and future scenes into a single experience can prompt viewers to question linear time intuitively. Literature can weave narratives that introduce block universe themes subtly, guiding readers to internalize complex ideas.

Educational curricula might evolve to teach temporal concepts not only as historical progressions or linear cause-and-effect sequences but as potentially more fluid constructs. Integrating lessons from cognitive science and physics into a single narrative could produce a generation comfortable grappling with non-linear time. Students might learn to question assumptions about causality, recognize that the arrow of time is not absolute, and appreciate the role of observers in shaping perceived reality.

Such cultural engagement ensures that if future experiments or theoretical successes bring time manipulation into the realm of the possible, societies will be better prepared, intellectually, ethically, and emotionally, to confront the implications. For example, a museum exhibit might allow visitors to experience a simulated block universe, fostering a deeper understanding of time's complexity through interactive displays.

Moreover, fostering public interest and understanding through media and public talks can bridge the gap between complex scientific theories and everyday experiences. Documentaries, podcasts, and online platforms can disseminate cutting-edge research in an engaging manner, sparking curiosity and encouraging informed discussions about the nature of time and reality.

**Preparing for the Unexpected**

One hallmark of cutting-edge research is the element of surprise. Remaining open to discoveries that challenge current assumptions is crucial. Future experiments might reveal that attempts to harness exotic matter always fail, or that consciousness studies show no fundamental role for quantum effects. Alternatively, novel phenomena could emerge, such as strange correlations across time in quantum experiments or unexpected stability in certain gravitational field configurations, forcing a revision of key theories.

Embracing the unexpected means fostering a research culture that values interdisciplinary dialogue, replication of results, and a humble acknowledgment of what we do not yet understand. If the first hint of temporal engineering appears, perhaps anomalous data in gravitational wave detectors or unprecedented outcomes in quantum simulation experiments, researchers must be ready to pivot, formulate new hypotheses, and test them rigorously.

Imagine a physicist stumbling upon unexpected data in a gravitational wave detector, hinting at previously unknown spacetime distortions. Instead of dismissing the anomaly, the scientific community would rally to explore its implications, potentially unlocking new aspects of the block universe and time manipulation theories. This proactive approach ensures that the field remains dynamic and adaptable, capable of integrating new information and adjusting theoretical frameworks accordingly.

Furthermore, establishing flexible research frameworks that can quickly incorporate new discoveries and shift focus when necessary will be essential. Creating environments where unconventional ideas are welcomed and tested without bias can foster innovation and accelerate breakthroughs.

**Collaborative Infrastructure and Funding**

Advancing research in these complex and interwoven fields requires robust infrastructure and sustained funding. Collaborative research centers that bring together experts from diverse disciplines can facilitate the exchange of ideas and resources. Securing funding from governmental agencies, private institutions, and international organizations is vital to support large-scale projects, cutting-edge experiments, and interdisciplinary studies.

Building international collaborations can pool expertise and resources, fostering a global approach to understanding spacetime and consciousness. Sharing data, tools, and findings across borders can enhance the quality and scope of research, ensuring that breakthroughs are accessible and can be built upon by the global scientific community.

Moreover, investing in education and training programs that prepare the next generation of researchers to work across disciplines will ensure sustained progress. Encouraging young scientists and scholars to engage in interdisciplinary studies can cultivate innovative thinking and equip them with the skills needed to tackle the multifaceted challenges ahead.

**Leveraging Big Data and Machine Learning**

The increasing availability of big data and advancements in machine learning present new opportunities for research. Analyzing vast datasets from astronomical observations, quantum experiments, and brain imaging studies can uncover patterns and correlations that might be missed by traditional analysis methods. Machine learning algorithms can identify subtle trends and make predictions, aiding in the discovery of new phenomena and the refinement of existing theories.

Imagine a machine learning model trained to detect anomalies in gravitational wave data, identifying potential signatures of exotic spacetime configurations that warrant further investigation. Similarly, AI-driven analyses of neural data could reveal how different brain regions interact to create the perception of time flow, providing insights into the cognitive processes underlying temporal experience.

Furthermore, machine learning can assist in optimizing experimental designs, managing large-scale simulations, and automating data analysis, allowing researchers to focus on interpreting results and developing new hypotheses. Integrating AI into research workflows can accelerate the pace of discovery and enhance the precision of scientific investigations.

**Engaging the Public and Fostering Scientific Literacy**

Public engagement and scientific literacy are essential for the continued support and advancement of complex research areas like spacetime and consciousness. Educating the public about the significance of this research and its potential implications can garner broader support and inspire future generations of scientists and thinkers.

Hosting public lectures, interactive workshops, and science fairs can demystify these concepts, making them more accessible and relatable. Creating educational materials that explain the block universe and simulation hypothesis in simple terms can help bridge the gap between specialized research and public understanding.

Moreover, fostering a dialogue between scientists and the public can lead to valuable feedback and new perspectives, enriching the research process. Encouraging citizen science initiatives where the public can participate in data collection and analysis can also enhance engagement and contribute to the advancement of knowledge.

**Ethical Frameworks and Responsible Research**

As research progresses towards potentially transformative technologies and deeper understandings of consciousness and spacetime, establishing robust ethical frameworks is paramount. Addressing the moral implications of manipulating time and space, altering consciousness, and simulating realities requires proactive consideration and regulation.

Developing guidelines that ensure the responsible use of temporal technologies can prevent misuse and protect individual and societal well-being. Ethical considerations should encompass issues such as privacy, consent, and the potential impact on mental health when exploring technologies that alter temporal perception or enable time manipulation.

Engaging ethicists and philosophers in the research process can help anticipate and address ethical dilemmas before they arise. Creating interdisciplinary ethics committees within research institutions can provide oversight and guidance, ensuring that advancements align with societal values and moral principles.

**Integrating Cultural and Historical Perspectives**

Understanding how different cultures and historical contexts have perceived and conceptualized time can offer valuable insights into contemporary scientific inquiries. Studying ancient myths, philosophies, and artistic expressions related to time can uncover underlying themes and intuitions that resonate with modern theories like the block universe and simulation hypothesis.

Exploring how various cultures have envisioned cyclical versus linear time can inform interdisciplinary research, highlighting how human perceptions of time are influenced by cultural narratives. This integration of cultural and historical perspectives can enrich scientific models, making them more comprehensive and reflective of the diverse ways humans relate to time.

Recognizing the cultural dimensions of temporal perception can guide the development of technologies and theories that are sensitive to different worldviews, fostering inclusivity and broadening the applicability of research findings.

## Expanding Educational Curricula

Educational institutions play a crucial role in shaping the future of research by equipping students with the knowledge and skills needed to navigate complex interdisciplinary fields. Expanding curricula to include integrated studies of physics, neuroscience, cognitive science, and philosophy can prepare students to tackle the multifaceted challenges associated with understanding spacetime and consciousness.

Incorporating hands-on projects, collaborative research opportunities, and interdisciplinary seminars can foster critical thinking and innovation. Encouraging students to engage in cross-disciplinary research can cultivate a generation of scientists and scholars adept at bridging gaps between fields, driving forward the exploration of time and consciousness.

Promoting STEM (Science, Technology, Engineering, and Mathematics) education alongside humanities can create well-rounded individuals capable of approaching problems from multiple perspectives, enhancing the depth and breadth of future research endeavors.

## Leveraging International Collaboration and Knowledge Sharing

Global collaboration is essential for addressing the grand challenges associated with understanding spacetime and consciousness. Sharing knowledge, resources, and expertise across international borders can accelerate progress and foster a more unified scientific community.

Establishing international research consortia focused on specific aspects of spacetime and consciousness can pool resources and expertise, enabling large-scale projects that individual institutions might find difficult to undertake alone. These collaborations can facilitate the exchange of ideas, standardize methodologies, and ensure that research findings are accessible to a broader audience.

Promoting inclusive and diverse research teams can enhance creativity and innovation, bringing together different cultural and intellectual perspectives that enrich the research process. International conferences, joint publications, and collaborative workshops can strengthen these global networks, fostering a sense of shared purpose and collective advancement.

**Utilizing Advanced Computational Tools and Simulations**

The complexity of block universe spacetime and consciousness requires advanced computational tools and simulations to model and understand their intricate dynamics. Developing sophisticated software and algorithms that can handle multidimensional data and simulate complex interactions is essential for advancing research.

Creating simulations that model the interaction between quantum fields and spacetime curvature can provide insights into how wormholes might form or stabilize. Similarly, developing neural network models that mimic the brain's temporal processing can help researchers understand how consciousness constructs the perception of time flow.

Integrating virtual reality (VR) and augmented reality (AR) technologies into research can provide immersive environments for testing theories and visualizing complex concepts. Imagine a VR simulation that allows scientists to navigate a block universe, observing how different observers perceive time from various spacetime slices. These tools can enhance both research and education, making abstract theories more tangible and comprehensible.

**Exploring the Role of Information Theory**

Information theory offers a powerful framework for understanding the fundamental nature of reality, particularly in the context of the simulation hypothesis and digital physics. Exploring how information is stored, transmitted, and processed within spacetime can shed light on the underlying structures that govern the universe.

Research into the informational basis of spacetime could reveal whether the fabric of reality is akin to a computational network, with information flowing and interacting in ways similar to data in a computer system. This perspective aligns with theories that view the universe as fundamentally digital, suggesting that spacetime is composed of discrete information packets rather than continuous fields.

Understanding the role of information in consciousness, how the brain processes and integrates information to create the perception of time, can bridge the gap between physical theories and cognitive experiences. This integration of information theory with neuroscience and physics can lead to a more unified understanding of how information underpins both the structure of spacetime and the workings of the human mind.

## Developing Ethical Guidelines for Block Navigation Technologies

As research progresses towards the possibility of block universe navigation and spacetime, developing ethical guidelines is imperative to ensure responsible use and prevent potential misuse. Establishing protocols that govern experimentation with temporal technologies can safeguard against unintended consequences and ethical breaches.

These guidelines should address issues such as:

<u>Consent and Autonomy</u>: Ensuring that individuals are fully informed and consent to participation in experiments involving temporal manipulation.

<u>Privacy</u>: Protecting the privacy of individuals in studies that may involve altering temporal perceptions or memories.

<u>Safety</u>: Implementing stringent safety measures to prevent harm resulting from unintended side effects of temporal manipulation.

<u>Equity</u>: Ensuring that advancements in temporal technologies are accessible and do not exacerbate existing social inequalities.

Collaborating with ethicists, policymakers, and community representatives would be imperative to develop comprehensive ethical frameworks that guide the responsible development and application of time manipulation technologies.

## Integrating Insights from Complex Systems and Emergent Phenomena

The study of complex systems and emergent phenomena can provide valuable insights into the dynamics of spacetime and consciousness. Understanding how simple interactions at a fundamental level give rise to complex behaviors and structures can inform models of the block universe and consciousness.

Exploring how emergent properties arise in cellular automata can shed light on how spacetime might emerge from underlying informational structures. Similarly, studying emergent consciousness in artificial neural networks can offer parallels to how consciousness arises from the brain's intricate network of neurons.

By applying principles from complex systems theory, researchers can develop models that capture the nonlinear and dynamic nature of spacetime and consciousness, enhancing our ability to simulate and understand these phenomena.

## Fostering Open Science and Data Sharing

Promoting open science and data sharing is essential for accelerating research progress and fostering collaboration. Making research findings, data sets, and computational tools openly accessible allows scientists worldwide to build upon each other's work, validate results, and develop new theories.

Establishing repositories for temporal geometry data, spacetime simulations, and neural recordings can facilitate the exchange of information and resources. Additionally, adopting open-source software for simulations and data analysis can democratize access to advanced computational tools, enabling a wider range of researchers to contribute to the field.

Encouraging transparency in research methodologies and results can enhance the credibility and reproducibility of studies, strengthening the foundation upon which future discoveries are made.

**Encouraging Public-Private Partnerships**

Public-private partnerships can play a crucial role in advancing research on block universe spacetime and consciousness. Collaborations between academic institutions, government agencies, and private companies can pool resources, expertise, and funding, enabling large-scale projects that drive innovation and discovery.

For example, partnerships with technology companies specializing in quantum computing, AI, or VR can provide access to cutting-edge tools and platforms necessary for simulating complex spacetime configurations or modeling neural processes. These collaborations can also facilitate the translation of theoretical research into practical applications, bridging the gap between science and technology.

**Leveraging Crowdsourcing and Citizen Science**

Crowdsourcing and citizen science initiatives can complement traditional research methods, engaging the public in data collection, analysis, and hypothesis generation. By involving a broader audience, scientists can harness diverse perspectives and distributed computational power, accelerating the pace of discovery.

Citizen scientists could contribute to large-scale simulations of spacetime or participate in experiments that map temporal perception under varying conditions. Engaging the public in these endeavors can also enhance scientific literacy and foster a sense of shared purpose in unraveling the mysteries of time and consciousness.

Crowdsourcing can help identify novel patterns or anomalies in data sets that might be overlooked by individual researchers, providing new avenues for exploration and understanding.

**Exploring the Intersection of Biology and Spacetime**

The relationship between biology and spacetime is a largely unexplored frontier that holds intriguing possibilities. Investigating how biological processes interact with the fabric of spacetime could reveal new dimensions of understanding about both life and the universe.

For example, exploring whether biological systems can influence local spacetime curvature or investigating the role of time perception in evolutionary processes can provide unique insights. Research into how organisms adapt to different temporal environments, such as varying light cycles or simulated time distortions, can inform our understanding of the adaptability and resilience of life.

Additionally, studying extremophiles, organisms that thrive in extreme conditions, could uncover how life copes with and possibly manipulates spacetime under extraordinary circumstances, offering clues about the limits and potentials of biological systems in relation to the universe.

### Integrating Artificial Intelligence in Research

Artificial Intelligence (AI) has the potential to revolutionize research in spacetime and consciousness by automating complex data analysis, optimizing experimental designs, and even generating new hypotheses. AI-driven models can process vast amounts of data far more efficiently than human researchers, identifying patterns and correlations that might remain hidden.

For example, machine learning algorithms could analyze gravitational wave data to detect subtle signatures of exotic spacetime phenomena, while AI-based neural networks could model the brain's temporal processing mechanisms with unprecedented accuracy. Additionally, AI can assist in optimizing simulations of the block universe, making them more efficient and scalable.

Integrating AI with quantum computing could open new avenues for simulating and understanding the interplay between quantum mechanics and spacetime, providing deeper insights into the fundamental nature of reality.

### Expanding the Scope of Simulation Studies

Expanding the scope of simulation studies to encompass a broader range of spacetime and consciousness scenarios can enhance our understanding and uncover new possibilities. Developing comprehensive simulation environments that model various aspects of the block universe, including different spacetime geometries, quantum field interactions, and cognitive processes, can provide a holistic view of how these elements interact.

Imagine a simulation platform where researchers can manipulate variables related to spacetime curvature, energy distributions, and neural activity, observing how these changes affect the overall system. Such an integrated approach can facilitate the exploration of complex interactions and emergent phenomena, providing a more nuanced understanding of the block universe and its implications for consciousness.

Additionally, creating collaborative simulation projects where researchers from different disciplines contribute their expertise can lead to more robust and versatile models, capable of addressing a wide array of research questions.

**Addressing the Challenges of Scalability and Complexity**

One of the significant challenges in researching spacetime and consciousness is managing the scalability and complexity of the systems involved. The block universe and conscious experience are inherently complex, requiring sophisticated models and computational resources to study effectively.

Developing scalable algorithms and leveraging high-performance computing can help manage this complexity, enabling researchers to simulate large-scale spacetime structures and intricate neural networks with greater accuracy and efficiency. Additionally, modular approaches that break down complex systems into manageable components can facilitate the study of individual elements while maintaining an integrated perspective.

Investing in computational infrastructure and fostering collaborations with computer scientists can provide the necessary tools and expertise to tackle these challenges, ensuring that research remains feasible and productive despite the increasing complexity of the questions being addressed.

## Exploring the Philosophical Dimensions of Temporal Manipulation

Temporal manipulation raises profound philosophical questions about the nature of reality, identity, and existence. Exploring these dimensions is essential for developing a comprehensive understanding of the implications of time manipulation technologies and theories.

Philosophers can engage with scientists to examine concepts such as causality, determinism, and the nature of change in the context of a manipulable block universe. Debating whether altering temporal structures could impact personal identity or the continuity of consciousness can provide valuable insights into the ethical and metaphysical ramifications of such technologies.

Moreover, exploring the subjective experience of time manipulation, how individuals perceive and adapt to changes in temporal flow, can bridge the gap between objective theories and human experience, enriching both philosophical discourse and scientific inquiry.

## Enhancing Simulation Fidelity and Realism

Enhancing the fidelity and realism of simulations is crucial for ensuring that they accurately reflect theoretical models and can provide meaningful insights. Developing more detailed and precise models of spacetime, quantum fields, and neural processes can improve the reliability and applicability of simulation results.

For example, incorporating higher-dimensional data, accounting for relativistic effects, and modeling complex interactions between particles and fields can enhance the realism of spacetime simulations. Similarly, integrating detailed neural architectures and cognitive processes into brain simulations can provide a more accurate representation of temporal perception and consciousness.

Further validating simulations against empirical data is essential for ensuring their accuracy. Comparing simulation outcomes with observational data from astronomical instruments, quantum experiments, and brain imaging studies can help calibrate models and refine their predictive capabilities.

**Encouraging Open-Source Research and Collaboration**

Promoting open-source research initiatives can accelerate progress by fostering collaboration and transparency. Making research tools, simulation software, and data sets freely available allows scientists worldwide to contribute to and build upon each other's work.

Open-source platforms can facilitate collaborative projects where researchers from different disciplines work together, sharing resources and expertise to tackle complex questions related to spacetime and consciousness. Additionally, open access to research findings can democratize knowledge, enabling a broader range of scientists and scholars to engage with and contribute to the field.

Encouraging the development of open-source frameworks for simulating spacetime and modeling consciousness can enhance reproducibility and innovation, ensuring that research advances are accessible and can be leveraged by the global scientific community.

## Investigating the Role of Entropy and Thermodynamics

Entropy and thermodynamics play a crucial role in our understanding of the arrow of time and the block universe. Investigating how entropy interacts with spacetime geometry and consciousness can provide deeper insights into the nature of temporal progression and the emergence of the subjective arrow of time.

Future research could explore how local decreases in entropy might influence spacetime curvature or how entropic processes are integral to the brain's temporal processing mechanisms. Understanding the relationship between entropy, information theory, and consciousness could uncover fundamental principles governing the flow of time and the structure of the block universe.

Examining the thermodynamic properties of hypothetical time manipulation technologies could inform their feasibility and potential impact on the broader spacetime fabric. This intersection of thermodynamics and spacetime research can lead to a more comprehensive understanding of the physical and informational underpinnings of temporal phenomena.

## Exploring Consciousness Beyond the Human Brain

While much of the focus is on human consciousness, exploring consciousness in other systems, biological or artificial, can expand our understanding of its relationship with spacetime. Studying how different organisms perceive and navigate time can reveal universal principles and unique adaptations, offering a broader perspective on the cognitive aspects of temporal experience.

Investigating artificial consciousness, whether through advanced AI or bioengineered systems, can provide insights into how consciousness interacts with spacetime in non-biological entities. Understanding the temporal processing capabilities of artificial systems can inform theories about the flexibility and scalability of consciousness in relation to the block universe.

Additionally exploring the possibility of collective consciousness or distributed cognitive systems can challenge traditional notions of individual temporal perception, prompting new theories about how groups of conscious entities interact with and perceive spacetime.

**Integrating Virtual and Augmented Reality in Research**

Virtual Reality (VR) and Augmented Reality (AR) technologies offer powerful tools for visualizing and experimenting with complex spacetime concepts. Integrating these technologies into research can enhance both the study and teaching of spacetime and consciousness by providing immersive environments that make abstract theories more tangible.

For instance, VR simulations can allow researchers to navigate virtual block universes, observing how different observers perceive time from various spacetime slices. These immersive experiences can aid in formulating and testing hypotheses about observer-dependent temporal perception and the dynamics of the block universe.

Similarly, AR can overlay complex spacetime data onto the physical world, enabling real-time visualization and manipulation of theoretical models. This can facilitate collaborative research efforts, allowing scientists from different disciplines to interact with and analyze data in a shared virtual space.

Incorporating VR and AR into educational programs can enhance learning by providing students with interactive and experiential understanding of complex spacetime and consciousness concepts, making the material more engaging and comprehensible.

**Utilizing Blockchain and Distributed Ledger Technologies**

Blockchain and distributed ledger technologies (DLTs) offer unique advantages for managing and securing vast amounts of research data related to spacetime and consciousness. These technologies can ensure the integrity, transparency, and accessibility of data, facilitating collaboration and trust among researchers.

For example, using blockchain to timestamp and verify simulation results can prevent data tampering and ensure that findings are credible and reproducible. Distributed ledgers can also enable decentralized data storage and sharing, allowing researchers worldwide to access and contribute to collective datasets without relying on centralized repositories.

Smart contracts, self-executing contracts with the terms directly written into code, can automate certain aspects of research collaboration, such as data sharing agreements, intellectual property rights, and funding distribution. This can streamline collaborative efforts and reduce administrative burdens, allowing researchers to focus more on their scientific inquiries.

## Exploring the Intersection of Consciousness and Information Processing

Consciousness and information processing are deeply intertwined, with information theory providing a framework for understanding how consciousness emerges from complex neural interactions. Exploring this intersection can shed light on how information is encoded, transmitted, and integrated within the brain, influencing our perception of time and the block universe.

Research could investigate how information flows through neural networks, how information compression and expansion relate to memory and anticipation, and how information processing dynamics shape temporal perception. Understanding these processes can inform theories about how consciousness constructs the experience of time and navigates the block universe.

Exploring the role of information in consciousness could lead to breakthroughs in understanding how information processing relates to spacetime geometry, potentially uncovering new principles that govern both cognitive and physical phenomena.

## Enhancing Collaboration Between Academia and Industry

Strengthening collaboration between academic institutions and industry can accelerate research advancements in spacetime and consciousness. Industry partners, particularly those in technology sectors like AI, quantum computing, and materials science, can provide valuable resources, expertise, and funding for cutting-edge research projects.

For example, partnerships with tech companies specializing in quantum computing can offer access to advanced computational resources necessary for simulating complex spacetime configurations or modeling neural processes. Collaborations with AI firms can facilitate the development of sophisticated algorithms for data analysis and pattern recognition in large datasets.

Additionally, engaging with startups and innovation hubs can foster a culture of experimentation and rapid prototyping, allowing researchers to test and iterate on theories and technologies more efficiently. These collaborations can bridge the gap between theoretical research and practical applications, driving forward both scientific understanding and technological innovation.

## Investigating the Role of Entanglement in Spacetime Dynamics

Quantum entanglement, the phenomenon where particles become interconnected and the state of one instantly influences the state of another, regardless of distance, has profound implications for our understanding of spacetime and consciousness. Investigating how entanglement interacts with spacetime geometry can uncover new dimensions of the block universe and its underlying informational structure.

Future research could explore whether entangled states can influence spacetime curvature or whether spacetime itself facilitates entanglement in ways not yet understood. Understanding the relationship between entanglement and spacetime could lead to breakthroughs in quantum gravity theories, providing a deeper understanding of how information is woven into the fabric of the universe.

Moreover, studying entanglement in neural systems could offer insights into how information is processed and integrated within the brain, potentially revealing new aspects of how consciousness emerges and operates within the block universe.

## Exploring Temporal Cognition in Artificial Systems

As artificial intelligence and robotics advance, exploring temporal cognition in artificial systems can enhance our understanding of consciousness and spacetime. Developing AI that can perceive, interpret, and manipulate temporal information in ways similar to humans can provide valuable models for studying the neural and cognitive aspects of temporal perception.

Creating AI systems capable of simulating time navigation within a block universe framework can offer insights into how temporal information is processed and utilized. These systems can be used to test theories about temporal perception, memory integration, and anticipation, providing a controlled environment for exploring complex cognitive phenomena.

Furthermore, integrating temporal cognition capabilities into robotics can enable more sophisticated interactions with dynamic environments, enhancing the utility and adaptability of robotic systems in real-world applications.

## Enhancing Public Understanding Through Media and Outreach

Enhancing public understanding of complex scientific concepts related to spacetime and consciousness is essential for fostering a supportive and informed society. Utilizing media and outreach programs can bridge the gap between specialized research and public knowledge, making these ideas more accessible and engaging.

Creating documentaries, podcasts, and online content that explain the block universe and simulation hypothesis in relatable terms can spark public interest and curiosity. Hosting public lectures, science fairs, and interactive workshops can provide opportunities for individuals to engage directly with researchers and ask questions about these intriguing concepts.

Additionally, developing educational resources that incorporate multimedia elements, such as animations, interactive simulations, and visualizations, can enhance comprehension and retention of complex ideas. By making these concepts more tangible and relatable, researchers can cultivate a more scientifically literate and engaged public, supporting the advancement of research through increased awareness and interest.

**Fostering a Culture of Curiosity and Open-Mindedness**

Cultivating a culture of curiosity and open-mindedness is fundamental for advancing research in spacetime and consciousness. Encouraging scientists and scholars to explore unconventional ideas, challenge existing paradigms, and embrace uncertainty can drive innovation and discovery.

Promoting interdisciplinary education and collaboration can foster an environment where diverse perspectives are valued and integrated, enhancing the depth and breadth of research. Creating safe spaces for brainstorming and idea exchange, where unconventional theories can be proposed and tested without fear of ridicule, can lead to breakthroughs that might otherwise remain unexplored.

## Conclusion

Future block universe timespace research directions span a broad spectrum, from refining theoretical physics to developing experimental methods, from understanding the cognitive underpinnings of temporal perception to fostering interdisciplinary collaborations that bridge diverse fields. These efforts are not just academic or intellectual pursuits, they are also cultural and existential endeavors that shape our understanding of reality and our place within it.

As we look to the future, embracing collaboration across disciplines, investing in technological innovations, and engaging with philosophical and ethical considerations will be paramount. Whether we ultimately discover ways to manipulate timelines or simply deepen our appreciation of how human minds navigate them, the ongoing exploration will enrich our understanding of reality itself.

By acknowledging the complexity and working collaboratively, the scientific community increases its chances of making genuine progress. The quest to unravel the mysteries of time, space, and consciousness is far from complete, but each step forward brings us closer to a more profound comprehension of the universe and ourselves.

# Chapter 17

## Conclusion and the Grand Speculative Leap

Contemplating the block universe, a realm where all moments, past, present, and future, coexist as a single, four-dimensional structure, can feel as alien as imagining a new color. For many, time is experienced linearly: one event follows another, and the past feels gone while the future remains unwritten. Yet physics challenges this intuitive understanding. The block universe, as suggested by relativity, presents time as a dimension akin to space. Every event, like every location, exists simultaneously in the grand tapestry of spacetime. What we experience as the "flow" of time might be better understood as our traversal through this unchanging structure, guided by the workings of memory, perception, and expectation.

From this perspective, I've ventured into profound questions: If time is not truly linear, could advanced technologies or intelligences traverse this manifold to visit the other coordinates within the block universe at will? The notion of navigating the block universe brings to mind immense scientific and philosophical challenges, ranging from the feasibility of wormhole stabilization to the ethical dilemmas of timeline alteration. However, while theoretical physics opens these doors conceptually, the practical barriers are staggering.

The mathematics of spacetime suggests that phenomena like negative energy and exotic matter are theoretically necessary to manipulate time. Yet these constructs remain elusive, existing only in abstract equations. Even if these materials could be harnessed, stabilizing wormholes or creating closed timelike curves would require energy levels far beyond our current technological grasp. Thus, while navigating the block universe is not forbidden by the laws of physics, it stands at the frontier of imagination and possibility.

## Speculative Scenarios: Agents of Change

Imagining the implications of time navigation reveals both its potential and its perils. Could secretive agencies or governments already be exploring this territory, even at the fringes? Speculation about unidentified aerial phenomena (UAPs) as experimental vehicles designed to probe spacetime has captured the imagination of many. Although no conclusive evidence supports these claims, the possibility invites us to consider a world where the manipulation of time becomes a geopolitical tool.

If governments could influence timelines, the consequences would be profound. Strategic interventions might reshape history to prevent wars or economic collapses. Conversely, such power could serve less noble purposes, entrenching inequality or erasing inconvenient truths. The absence of accountability in such scenarios underscores the ethical gravity of temporal manipulation. Would humanity, as it exists today, be capable of wielding such power responsibly?

## The Role of Consciousness: Architects of Experience

Beyond engineering challenges, the block universe poses philosophical questions about the role of consciousness in shaping temporal experience. While time may be static in the block universe, we perceive it dynamically, as an unfolding narrative. This perception is rooted in how our brains process information, integrating sensory input, storing memories, and constructing a sense of continuity.

Consciousness might play a more profound role than simply interpreting the block universe. Could it be that the mind itself imposes an order on events, selecting "now" from the manifold of spacetime? If so, consciousness becomes not just an observer but a participant in the unfolding of time. Advances in neuroscience and cognitive science are already shedding light on how humans perceive duration and sequence, offering insights that might one day bridge subjective experience with the physics of time.

**Ethical and Philosophical Considerations**

If time navigation becomes a reality, it will demand a framework of ethical principles to guide its use. Should humanity have the right to navigate timespace, knowing that even mere observation might ripple unpredictably across the block universe? Would erasing tragedies also erase the growth and resilience forged in their wake? Ethical questions of this magnitude challenge us to consider the nature of free will, identity, and justice in a world where the past and future can be shaped as easily as the present.

One scenario to consider is the potential misuse of temporal power. Could governments or organizations use block universe navigation to suppress dissent, rewrite cultural histories, or engineer events that favor particular agendas? These possibilities, while speculative, remind us that the ethical dilemmas of today, privacy, inequality, and environmental stewardship, would only magnify in a world where time itself could be controlled.

Conversely, block universe navigation could offer unprecedented opportunities for learning and healing. Imagine historians able to directly observe pivotal events, correcting inaccuracies and deepening our understanding of the past. Medical teams might travel to the origin coordinates of pandemics, armed with advanced knowledge to prevent global crises. These optimistic visions illustrate how temporal technologies might serve as tools for progress, provided their use is tempered by wisdom and restraint.

**A Glimpse into the Distant Future**

Looking to far "future" coordinates, one can imagine humanity, or its descendants, transcending the limitations of "now". Suppose theoretical barriers are overcome, and practical solutions to block universe navigation are developed. What would such a world look like? Such travel might become as routine as air travel, governed by strict ethical codes and advanced oversight systems.

In this speculative scenario, history would no longer be a fixed record but a resource to be explored, understood, and, perhaps, selectively adjusted. Diplomacy might extend across centuries, with leaders from different eras convening to resolve conflicts or align goals. Education could take on new dimensions, with students experiencing historical events firsthand rather than reading about them in books.

Yet, this vision is not without risks. Altering timelines could produce unintended consequences, fracturing reality into branching paths or multiverses. Attempts to rewrite history might encounter natural limits, as if the block universe resists change to preserve causal consistency. These speculative challenges highlight the need for humility as humanity advances toward unprecedented capabilities.

## Consciousness as a Pathway

If consciousness is integral to experiencing time, then block universe navigation might not rely solely on physical engineering. Advanced beings might refine their cognitive architectures, altering their perception to "tune into" different epochs without moving physically through spacetime. This approach blurs the line between external navigation and internal transformation, suggesting that the key to such navigation might lie within the mind itself.

This concept aligns with philosophical traditions that view consciousness as a bridge between subjective experience and objective reality. If advanced civilizations discover ways to reconfigure their awareness, they might access the block universe in ways that transcend physical constraints.

## Cultural and Existential Implications

The block universe compels us to reconsider what it means to exist. If all moments are equally real, then concepts like "progress" and "regret" take on new meanings. Tragedies remain fixed in the timeline, yet so do triumphs. The human experience, marked by change and growth, gains a new dimension when viewed against the backdrop of an unchanging spacetime.

## The Enduring Mystery

As I conclude this exploration, I am struck by the duality of the block universe. It is most certainly a mathematical construct and a profound mystery. Its elegance lies in its infinite complexity: all moments existing, unchanging, in a grand tableau. Yet its implications challenge our deepest intuitions about free will, causality, identity, and the nature of existence.

Whether the block universe remains an enduring mystery or becomes a tangible reality, the journey to understand timespace will continue to enrich humanity. Each new discovery brings us closer to comprehending not only the universe but also our place within it.

The block universe dares us to think expansively, to question assumptions, think out of the box, and to embolden our quest, to know the unknown.

## **References**

Alcubierre, M. (1994). The warp drive: hyper-fast travel within general relativity. Classical and Quantum Gravity, 11(5), L73–L77.

Beane, S. R., Davoudi, Z., & Savage, M. J. (2012). Constraints on the universe as a numerical simulation. European Physical Journal A, 50, 148.

Bostrom, N. (2003). Are you living in a computer simulation? The Philosophical Quarterly, 53(211), 243–255.

Casimir, H. B. G. (1948). On the Attraction Between Two Perfectly Conducting Plates. Proceedings of the Koninklijke Nederlandse Akademie van Wetenschappen B, 51, 793–795.

Chalmers, D. J. (2005). The Matrix as metaphysics. In C. Grau (Ed.), Philosophers Explore The Matrix (pp. 132–176). Oxford University Press.

Eagleman, D. M. (2009). Human time perception and its illusions. Current Opinion in Neurobiology, 18(2), 131–136.

Einstein, A. (1905). On the Electrodynamics of Moving Bodies. Annalen der Physik.

Einstein, A. (1916). The Foundation of the General Theory of Relativity. Annalen der Physik.

Everett, H. (1957). 'Relative State' Formulation of Quantum Mechanics. Reviews of Modern Physics, 29, 454–462.

Fredkin, E. (2003). Digital Philosophy. International Journal of Theoretical Physics, 42(2), 189–247.

Gates, S. J. (2011, June). Symbols of power: Adinkras and the nature of reality. Physics World. Retrieved from https://physicsworld.com/

Gödel, K. (1949). An Example of a New Type of Cosmological Solutions of Einstein's Field Equations of Gravitation. Reviews of Modern Physics, 21, 447–450.

Greene, B. (2011). The Hidden Reality: Parallel Universes and the Deep Laws of the Cosmos. Alfred A. Knopf.

Hafele, J. C., & Keating, R. E. (1972). Around-the-World Atomic Clocks: Predicted Relativistic Time Gains. Science, 177(4044), 166–168.

Hawking, S. W. (1992). Chronology Protection Conjecture. Physical Review D, 46(2), 603–611.

Hooft, G. (2016). The Cellular Automaton Interpretation of Quantum Mechanics. Springer.

Lewis, D. (1976). The Paradoxes of Time Travel. American Philosophical Quarterly, 13, 145–152.

LIGO Scientific Collaboration. (2016). Observation of Gravitational Waves from a Binary Black Hole Merger. Physical Review Letters, 116(6), 061102.

Minkowski, H. (1909). Space and Time. Physikalische Zeitschrift, 10, 104–111.

Morris, M. S., & Thorne, K. S. (1988). Wormholes in spacetime and their use for interstellar travel: A tool for teaching general relativity. American Journal of Physics, 56, 395–412.

Putnam, H. (1967). Time and Physical Geometry. Journal of Philosophy, 64(8), 240–247.

Rich, B., & Janos, L. (1994). Skunk Works. Little, Brown & Company.

Schellekens, A. (2013). Life at the centre of the universe. Scientific American, December issue.

Thorne, K. S. (1994). Black Holes and Time Warps: Einstein's Outrageous Legacy. W. W. Norton & Company.

Virk, R. (2019). The Simulation Hypothesis: An MIT Computer Scientist Shows Why AI, Quantum Physics, and Eastern Mystics All Agree We Are in a Video Game. Bayview Books.

Wheeler, J. A. (1962). Curved Empty Space–Time as the Building Material of the Universe. Revista Brasileira de Física, Edição Comemorativa.

Wheeler, J. A. (1990). Information, physics, quantum: The search for links. In W. H. Zurek (Ed.), Complexity, Entropy, and the Physics of Information (pp. 3–28). Addison-Wesley.

Wheeler, J. A., & Feynman, R. P. (1945). Interaction with the Absorber as the Mechanism of Radiation. Reviews of Modern Physics, 17(2–3), 157–181.

www.ingramcontent.com/pod-product-compliance
Lightning Source LLC
Chambersburg PA
CBHW020635220526
45464CB00001B/163